SCIENCE 101

PHYSICS

HarperCollins books may be purchased for educational, business, or sales promotional use. For information, please write: Special Markets Department, HarperCollins Publishers, 10 East 53rd Street, New York, NY 10022.

Produced for HarperCollins by:

Hydra Publishing
129 Main Street
Irvington, NY 10533
www.hylaspublishing.com

FIRST EDITION

Library of Congress Cataloging-in-Publication Data has been applied for.

ISBN: 978-0-06-089134-3
ISBN-10: 0-06-089134-3

07 08 09 10 QW 10 9 8 7 6 5 4 3 2 1

SCIENCE 101
PHYSICS

Barry Parker

Collins
An Imprint of HarperCollins*Publishers*

CONTENTS

Frequency waveforms.

WELCOME TO PHYSICS

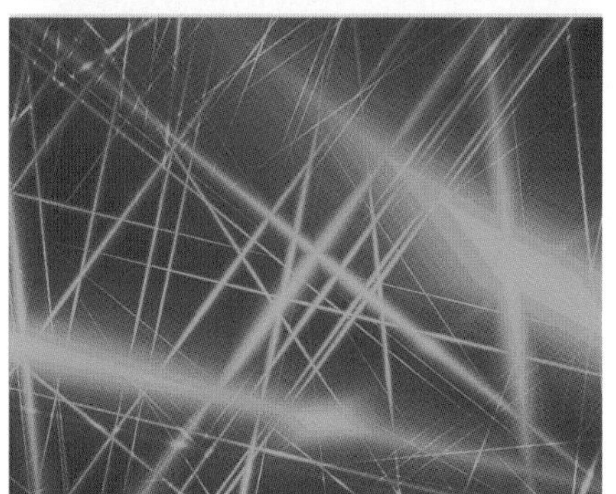

Left: Roller coaster. Top: Atom structure. Bottom: Field of blue lasers.

What is physics? The simplest answer is that physics is the science that deals with matter, energy, space, and time. These concepts are basic to all phenomena, so it is no exaggeration to say that physics is at the foundation of all science. Because of this, physics has been called the "king of sciences."

The aim of physics is to understand how nature works. In order to reach this understanding of the physical world, careful observation and experimentation are required, the results of which are analyzed. From this analysis, theories are developed. Predictions can be made from these theories that may lead to a refinement of a theory, or to serious conflict, in which case a theory has to be rejected. Part of this theorizing involves the search for basic laws of the universe. One of the key ideas of physics is that behind the apparent complexity of the world there is an underlying simplicity and unity.

WHY IS PHYSICS IMPORTANT?

Physics is integral to our society. It generates the knowledge needed for technology, but it is also an enterprise that expands our horizons and pushes back the frontiers of our knowledge. It plays a vital role in the education of many types of scientists—engineers, chemists, meteorologists, astronomers, and many types of medical experts. Physics has also provided us with many comforts and has helped extend our lives through its many important contributions to medicine.

Physics is critical to our future. It will be needed to help solve many of the problems now facing the world, such as global warming, waning energy resources, and the poisoning of our atmosphere.

It might seem that physics is a closed science in that most of the major discoveries are behind us, but this pronouncement has been made several times throughout history. As interest in the field waned at the end of the nineteenth century, no one could have imagined that in the next few years some of the most important discoveries in physics would be made, including relativity and quantum theory.

THE MANY BRANCHES OF PHYSICS

Physics is divided into sections that deal with separate phenomena. Mechanics, developed by Isaac Newton in the seventeenth century, is one of the oldest branches. It deals with motion, force, energy, and inertia, and how these things act on solids and fluids.

The study of heat is concerned with the principles of temperature measurement, heat flow, and the effect of heat on the properties of matter. Closely related to heat is thermodynamics, the study of the effects

Ice melting.

Two burners on a natural gas kitchen stove.

of change of temperature, pressure, and volume on physical systems.

The phenomenon of waves plays an important role all through physics. Sound is a wave, and in the nineteenth century it was found that electricity and magnetism together create electromagnetic waves. Light is an electromagnetic wave, as are radio waves, ultraviolet, infrared, and X-rays.

The study of light led to the field of optics, which is concerned with the nature and propagation of light and its refraction through various transparent media. Light can also be separated into various frequencies, forming what are known as spectra. The understanding of spectra provided a tremendous amount of information about the elements, the stars, and other celestial bodies.

One of the most fascinating areas of physics is referred to as modern physics. This has several subdivisions, including atomic physics, quantum mechanics, relativity theory, nuclear physics, and solid state physics. Quantum mechanics was developed in the late 1920s and is fundamental to all physics. It tells us how electrons move around the nucleus of the atom and what happens when they jump between orbits.

Relativity shows that mass and energy are equivalent and explains the source of energy for the Sun and stars. Relativity has also given us a new insight into gravity. Newton described gravity as force at a distance, but according to relativity it is curved space. Relativity has also enabled us to understand neutron stars and black holes, giving us profound insight into the workings of the universe.

When basic physics is applied to other disciplines, the result is a number of subfields, such as biophysics, geophysics, astrophysics, and elementary particle physics. The physics of the Earth, namely geophysics, atmospheric physics, and physical oceanography, are also important subfields.

From classical mechanics to quantum mechanics, *Science 101: Physics* delves into the laws of the universe and follows the journey of the brilliant people who were driven to understand the world around them.

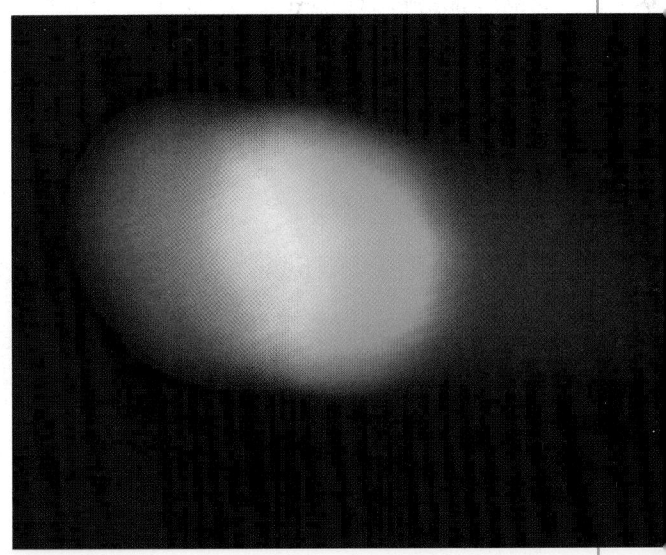

Light spectrum.

CHAPTER 1

NEWTON, MOTION, AND CLASSICAL MECHANICS

Left: The acrobatic movements of a skateboarder are captured with high-speed photography. The nature of motion was considered for thousands of years before Newton devised his groundbreaking theories on this complex concept. Top and bottom: Motion is prevalent throughout nature, as seen in the movement of water in a geyser and in a waterfall.

The early Greek philosophers made several important advances in our understanding of the natural world, but most of what they did was abstract philosophizing. They did not believe in experimentation and felt that all answers could be found through pure reason. It was not until the sixteenth century that someone decided to test the ideas and assumptions that had to that point been accepted as fact. Galileo Galilei (1564–1642) realized that all beliefs about our world had to be proven before they could be accepted. He did not accept, for example, the idea that heavy objects fell faster than light ones, and experimentation soon showed him he was right. Galileo started a revolution in physical science, but it was Sir Isaac Newton (1642–1727) of England who brought it to fruition and put it on a firm foundation.

Newton's intellect changed science forever and gave us what we call classical mechanics, the branch of physics that studies the deterministic motion of objects.

Early Ideas

Many early scholars no doubt wondered about the physical phenomenon we know as motion. Fortunately, some of the records from these Greek philosophers have survived, so we know their thoughts on many types of motion: the motion of stones that were dropped or thrown, the motion of objects that were pushed or pulled, the motion of flames and smoke, and the motion of bubbles in water, to name a few. One of the most influential of these early philosophers was Aristotle, who lived from 384 to 322 BCE.

Above: Aristotle believed that everything was made of four elements: water, fire, earth, and air. Top left: The Greek philosopher Aristotle. His thoughts on motion remained unchallenged until Galileo questioned them in the sixteenth century.

ARISTOTLE

Aristotle believed that the Earth was made up of four elements—earth, water, air, and fire—and that each of these elements had its "natural" place according to its weight. The element earth was heaviest, so it was closest to the center of the planet. Directly above earth was water, then air, and finally at the outer edge was fire.

Within this scheme, motion was also a natural phenomenon. If you raised a rock and dropped it, it fell through the air to its natural place, namely the ground (or Earth). It was understood, however, that a rock could be thrown upward, which did not appear to be a natural motion. Aristotle explained this by saying that the impulse given to the stone was transmitted to the air, and the air carried the stone. But it only carried it a small distance before natural motion took over and brought it back to Earth.

Some things had to rise to achieve their natural place. Bubbles rose through water, and flames rose through air. Aristotle also had an explanation for motion in the heavens. Heavenly objects such as the Sun, Moon, and stars were not composed of the same elements as those on Earth, so they did

Aristotle believed that everything had a natural place, and that when a stone was dropped it fell to its natural place, the Earth. He had problems, however, applying this idea to a stone thrown upward.

not behave in the same manner. These objects were composed of a fifth element called the fifth essence, or aether. The natural motion of heavenly objects was in circles around the center of the universe, and the Earth was the center of the universe. Some of the heavenly bodies, namely stars, appeared to be at rest. Aristotle had no explanation for these celestial objects.

GALILEO

No one challenged Aristotle's ideas of motion and the makeup of the Earth for almost 2,000 years. The first individual on record to test these concepts was the Italian physicist and mathematician Galileo Galilei. Born in Pisa, Italy, in 1564, Galileo was skeptical of many of Aristotle's teachings and eventually decided to prove them for himself.

Aristotle had put forward the idea that heavier objects fell at a faster rate than lighter objects. He pointed to leaves and snowflakes, comparing them to stones. Yet he never tested his assumptions. Galileo began by considering a cannonball. According to Aristotle, if you dropped the ball it would fall at one rate, but if you cut it in half the two halves would fall at much slower rates. Galileo was sure this was wrong. After a few simple experiments, he realized that air was playing a significant role and came to the conclusion that if the air was absent, everything would fall at the same rate. Unfortunately, he had no way of proving this, because he could not create a vacuum.

Italian philosopher, astronomer, and mathematician Galileo Galilei made fundamental contributions to the study of motion and other natural phenomena.

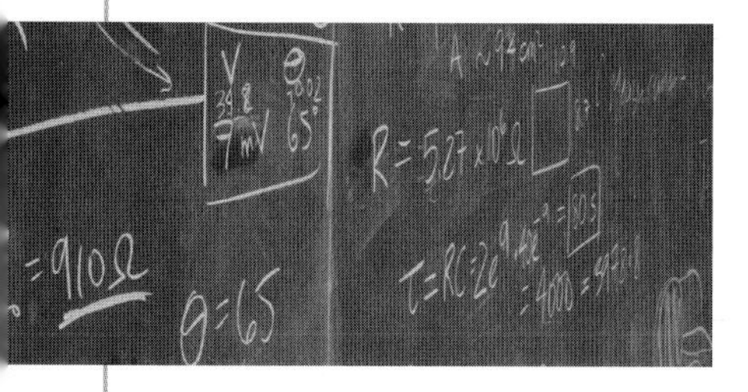

Sir Isaac Newton

Born prematurely in Woolsthorpe, England, on Christmas Day, 1642, Isaac Newton almost died at birth. His father had previously passed away, and when his mother remarried three years later he was sent to live with his grandparents. Young Isaac was not an outstanding student, preferring to spend his time building windmills and water clocks. His mother eventually decided to take him out of school so he could help on the family farm, but it was soon obvious he was not suited for this type of work. He was usually found reading under a tree when he was supposed to be working. Several of Isaac's teachers eventually persuaded his mother to send him to Trinity College at Cambridge.

Newton entered the college in 1660 and graduated in 1664. Little is known about his academic record, but he did attract the attention of one of his teachers, Isaac Barrow. In 1665 the plague arrived in England and Newton was forced to return to his mother's farm at Woolsthorpe. During this time he made several important discoveries, including the laws of motion and the law of gravity. It was also at Woolsthorpe that Newton developed calculus.

CALCULUS

One of Newton's most important contributions was the development of the powerful mathematical tool that he referred to as the "method of fluxions." It was important in dealing with instantaneous rates of change, and also in calculating areas and volumes. The same ideas in a slightly different form were discovered about the same time by Gottfried Leibniz in Germany. A controversy eventually arose when friends of Newton in England began claiming that Leibniz had stolen Newton's idea. There is little doubt that Newton came

Above: Sir Isaac Newton, who is now considered to be one of the greatest scientists of all time. Top left: One of the most important contributions that Newton made to science was the development of the so-called language of physics, calculus.

up with the idea first, but he was slower to publish it. Leibniz, who was a very able mathematician, no doubt came up with the idea independently and was the first to make it available to the scientific public. Furthermore, Leibniz used a better notation, similar to the one that is used today. Calculus eventually became an indispensable tool in science and engineering.

Isaac Newton's birthplace, Woolsthorpe Manor, in Lincolnshire, England.

IN HIS OWN WORDS

When he died, Newton was buried in Westminster Abbey along with many other revered and honored Englishmen. His legacy is indeed astounding: He made important contributions to mechanics, optics, mathematics, astronomy, and our understanding of gravity, revolutionizing physical science. How did he make these discoveries? Newton tried to explain his success as, in part, the result of the work of those who went before him, saying, "If I have seen farther than other men, it is because I stood on the shoulders of giants." Newton was also in awe of the world around him: "I do not know what I may appear to the world; but to myself I seem to have been only like a boy playing on the seashore, and diverting myself in now and then finding a prettier shell than ordinary, while the great ocean of truth lay all undiscovered before me."

THE *PRINCIPIA*

English astronomer Edmond Halley (1656–1742), after whom the famous comet is named, encouraged Newton to publish the many discoveries he had made. Newton finally agreed, and in 1687 he began writing them down. The book was called *Philosophiae Naturalis Principia Mathematica* (*Mathematical Principles of Natural Philosophy*), commonly referred to as simply the *Principia.* Originally written in Latin, it was difficult for most people to read, which was intentional. Newton had been criticized on some of his earlier discoveries by people who were ill-qualified to judge them, and he did not want it to happen again.

The *Principia* was considered one of the greatest books of its kind ever written, but Newton had to overcome some hurdles before he completed it. Robert Hooke, a natural scientist and Newton's bitter rival, claimed that some of the ideas put forth in the *Principia* were stolen from him. Newton was outraged and threatened to withdraw the book, but Halley quickly intervened and smoothed things over. The Royal Society, Britain's national academy of science and Newton's publisher, ran short of funds, so Halley stepped in and offered to pay for publication. Eventually, 2,500 copies were published, and scientists throughout Europe quickly saw it for the masterpiece it was.

Newton held several important positions during the later part of his life, but he made few discoveries during this time. He did publish one other groundbreaking book in 1704, titled *Opticks*, which contained his discoveries in light and optics. In 1689 Newton was elected Member of Parliament. In 1696 he was appointed ward of the British Royal Mint, and in 1697 he was promoted to master. In 1703 he was elected president of the Royal Society, a position he held until his death in 1727.

Motion

In today's world, motion takes many forms, ranging from the motion of bicycles to that of airplanes; from the motion of the tiny particles that make up the universe to that of the galaxies. In most cases motion is very complex, and in order to understand it we must begin with its simplest properties.

SPEED AND VELOCITY

The simplest type of motion is that of an object traveling at a uniform speed in a given direction. Consider, for example, a car traveling at a uniform speed of 50 miles per hour (mi/hr) along a straight highway. If we are in the car, we know we will travel 100 miles in two hours, 200 miles in four hours, and so on. But in practice, a car rarely stays at the same uniform speed for any length of time. For a while it might be traveling at 60 mi/hr, then a little later, at 40 mi/hr. Also, there will be turns in the road, so it is not always going in a straight line. Nevertheless, when we arrive at our destination we have traveled a certain distance in a certain time, and from this information we can calculate our average speed by dividing the distance traveled by the elapsed time. If it took us half an hour to go 27 miles, our average speed was 54 mi/hr, but we may have deviated considerably from it over the course of the journey. In dealing with cars, we are most familiar with the unit mi/hr, but there are several other units used when describing speed. Feet per second (ft/sec) is another common unit, in the mks (meter, kilogram, second) system of units we have meters per second (m/s), and in the cgs (centimeter, gram, second) system we have centimeters per second (cm/sec).

If during our trip we changed direction several times, it would be convenient to have some method of keeping track of these changes. To do this we use *velocity*. Velocity has both magnitude (speed) and direction. So when we talk about the velocity at a particular point or time in our trip we have to specify both the

Above: One of the easiest ways to understand speed, velocity, and acceleration is to relate them to a car in motion. Top left: A speedometer indicates the speed at which a car is traveling. This is measured for American automobiles as miles per hour.

speed of our car and the direction in which it is traveling.

In order to calculate our speed at any given time or any given point in a trip, we can use the same method as for average speed; all we have to do is make the elapsed time very short. In fact, if we want the instantaneous velocity, we have to make it infinitesimally short. This might appear to be a difficult task, but it is relatively easy to do using calculus.

ACCELERATION

During our trip we changed velocity a number of times—at one point we may have been going 60 mi/hr, but a few minutes later we slowed down to 45 mi/hr. These changes in velocity are referred to as *accelerations*. An increase in speed is commonly called acceleration, while a decrease is known as deceleration, but in fact a deceleration is nothing more than a negative acceleration. Acceleration is defined as change in velocity divided by elapsed time. As with velocity, if acceleration is calculated over a long period of time the result is average acceleration. More often, however, it is the measurement of *instantaneous* acceleration that is sought, so an infinitesimally short period of elapsed time is used in the calculation.

The units of acceleration are the units of velocity divided by time. This gives us feet per second per second (ft/sec^2), or m/s^2 in the mks system and cm/s^2 in the cgs system.

VECTORS AND SCALARS

Returning to our discussion of speed and velocity, we saw that they were different in that velocity had both magnitude and direction, whereas speed had only magnitude. We refer to a quantity that has both magnitude and direction as a *vector*; quantities that have only magnitude are called *scalars*. Examples of vectors are force, velocity, and momentum. Examples of scalars are temperature, energy, length, mass, and volume.

It is common practice to use arrows to designate vectors. The direction of the arrow represents the direction of the vector; its length represents its magnitude. As we will see, vectors are of considerable value to physicists and mathematicians, and they are used extensively.

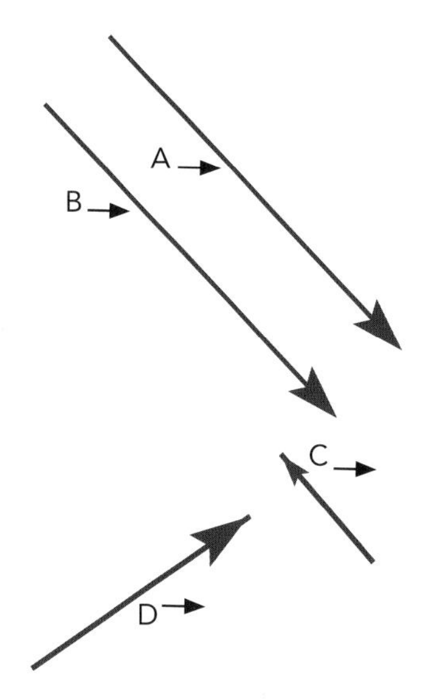

In the top figure, lines A and B represent equal vectors—two vectors that have equal magnitudes and the same direction. Vectors D and C are unequal.

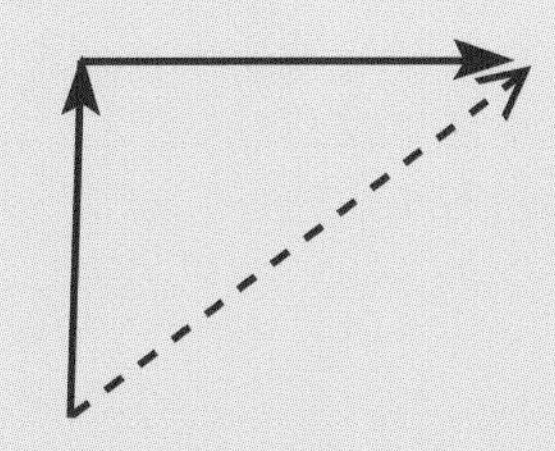

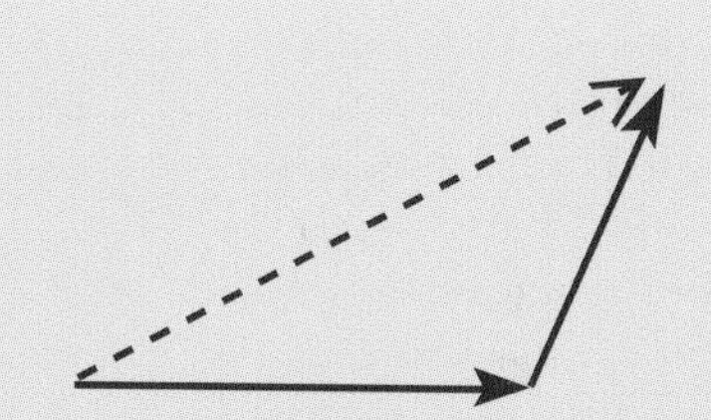

A resultant *is the name given to a vector that represents the sum of two or more vectors, in this case, the dotted lines.*

THE ADDITION OF VECTORS

Just as velocities are frequently added or subtracted when we are dealing with more than one, so too are vectors added and subtracted. The easiest case is where the vectors are in the same direction—they are simply added in the usual way and if they are in opposite directions they are subtracted. We may, however, have two vectors that are perpendicular to each other as in the left figure. In this case we have an action (or motion) to the north, then one to the east, and we end up at the end of the second vector. It's easy to see that we would have the same result if we took the dotted line to the same point, so our "resultant" (the addition of the two vectors) is an arrow represented by the dotted line. It is also possible to add vectors that are not perpendicular to each other as in the figure at right. Again the resultant vector is the dotted one. And we can, of course, extend this technique to several vectors.

Forces and Newton's First Law

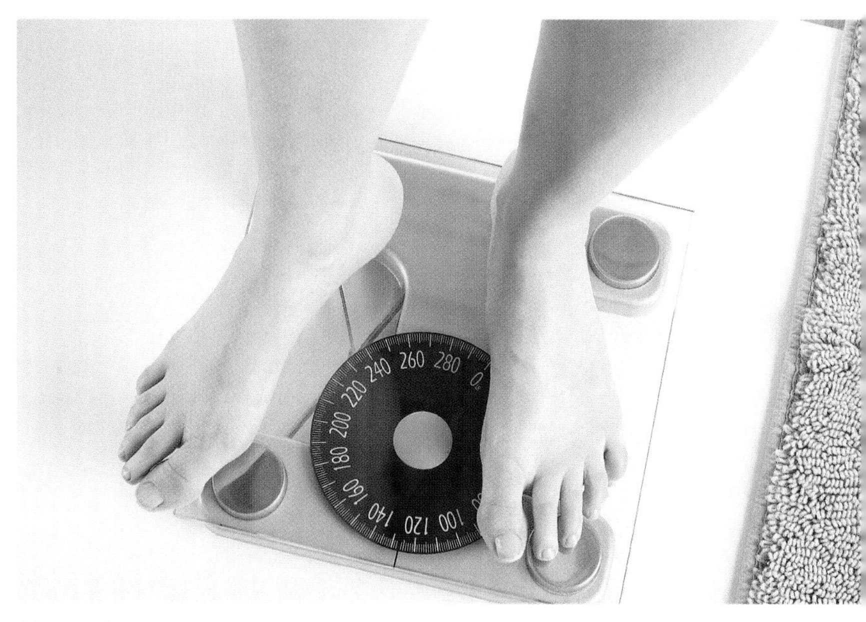

Above: When you stand on a scale you get a measure of your weight, which depends on the Earth's gravitational field. Your mass is independent of gravity. Top left: Pushing a car that is at rest requires a certain force, which will result in a change in velocity.

Up to this point we have described motion. The study of motion is called kinetics. Gravity is a force, and you are probably familiar with its units of measurement, which, in America, are pounds and ounces.

FORCE

In order to make an object move, you must push or pull on it. This push or pull is called a *force.* More specifically, a force imposes a change in velocity on an object—if the object is at rest it is set into motion; if it is already in motion with a certain velocity, the force changes its velocity. And since a change in velocity is the definition of acceleration, this tells us that a force is associated with acceleration. Like velocity, force is a vector. When we push on a cart, for example, we apply a force of a certain magnitude, and it is in a certain direction. For the units of force we have to define a new concept, called *mass.*

MASS AND WEIGHT

Everyone knows what weight is; all you have to do to determine your weight is step on a scale. What is actually being measured in this case is the gravitational pull of the Earth on your body. Measurement of weight depends on the gravitational field that an object is weighed in. If you went to Mars or Jupiter, for example, you would have a different weight, and in space (away from any gravitating object) you would weigh nothing.

What would be more useful than a weight measurement would be a measure of the amount of material in a body that is independent of the gravitational field. Such a measurement is referred to as *mass.* To get the mass of an object we divide the weight of the object by the magnitude of the gravitational field it is in. Our mass is therefore the same regardless of where we happen to be; it is the same on Earth, on Mars, or out in space.

The units of mass are kilograms (kg) in the mks system and grams (g) in the cgs system. In the British system they are referred to as slugs.

We can now define the units of force. Force is mass multiplied by acceleration, or, in mks units, $kg \times m/sec^2$, which is called a newton. In cgs units it is g × cm/sec^2, which is called a dyne. In the British system the unit of force is a pound.

INERTIA

Closely related to mass is what is called inertia. It is something we encounter every day. For example, if you try to push a stalled car, you know that you have to exert a force to get it moving. The tendency for something at rest or in uniform motion to remain in whatever state it is in is called inertia.

Galileo realized that inertia was an important property of matter. He knew it was a resistance to a change in motion, and he knew that heavy objects had more inertia than light ones, since it took a greater force to push a heavy object than it did a light one; but he was never able to fully understand the concept.

An ice skater is an example of something that slows down relatively slowly.

THE ADDITION OF FORCES

Forces are vectors that can be added or subtracted in the same way as velocity vectors. As an example consider a boy sitting on a swing that hangs from a rope. When the swing is pulled to the side, as in the diagram, several forces are involved. There is the weight of the boy and the swing, which is acting downward; the outward pull on the rope; and an upward force along the rope that holds the swing. We can represent these forces as shown in the diagram. For a condition in which there is no change in the motion of the system, called equilibrium, we have to have the vectors representing these forces forming a closed triangle, as shown in the diagram. Using this we can determine any of the forces if two are known.

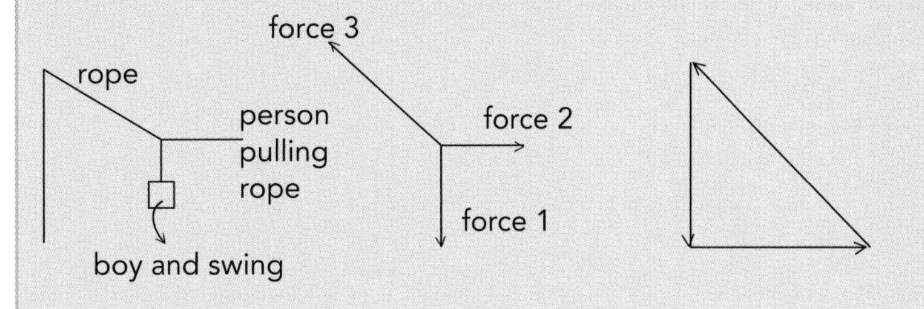

Several forces are involved in pulling a weight on a string to one side.

NEWTON'S FIRST LAW OF MOTION

Newton was the first to understand and describe inertia, and he incorporated it into what is known as his first law of motion:

A body will continue in a state of rest or uniform motion in a straight line, unless acted upon by a force.

This law, often referred to as the Law of Inertia, seems to defy common sense at first glance. Do objects in uniform motion actually continue in their motion indefinitely? If it were true and you were traveling 60 mi/hr in your car, when you took your foot off the gas you would continue going 60 mi/hr indefinitely, and you know that this is not the case. Rather, the car slows down almost immediately. This is because external forces such as friction and air pressure cause the car to slow and stop. Newton's law states that the body will continue in uniform motion if no external forces act on it. An example that comes close to illustrating the law is a skater gliding over the ice. When the skater no longer applies a thrusting force, she will continue to glide for a long distance before stopping. But even in this case there is a small amount of friction, which will eventually cause the skater to stop.

How is inertia related to Newton's first law? According to the principle of inertia, which was formulated by Galileo, a body moving uniformly on a level surface will continue in the same direction at the same constant speed unless acted upon by a force. Newton's first law, therefore, is a definition of the principle of inertia.

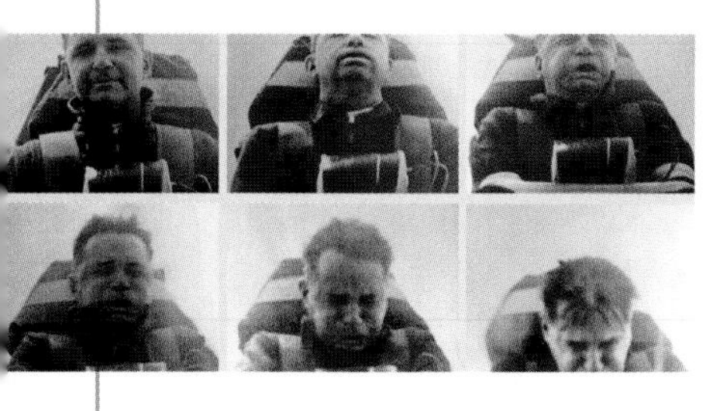

Acceleration and Newton's Second Law

So far, we know that forces and accelerations are related, that an object can have several forces acting on it, and that we can add and subtract these forces. Indeed, if the appropriate additions are done, we find that a single force can be used to represent the combined forces acting on an object, which is referred to as the net force.

To see how net force relates to acceleration, consider a simple experiment. Assume we have several blocks of different weights, and therefore different masses, on a smooth plane (by smooth we mean friction-free). We want to push or pull the blocks, so we will need a device with which to

measure the force we are exerting. A simple spring scale will do. If we pull on the lightest block, we will give it an acceleration, since it started at rest. Now pull the second-lightest block (which has a greater mass) with the same force using the scales. Again we see that we give the block an acceleration, but it is less this time. We can continue this with the other blocks, each heavier than the last, and we will see that in each case the acceleration will be less than the one before it. Why does this happen? This question is answered by Newton's second law of motion:

The acceleration produced by a force acting on a body is directly proportional to the magnitude of the force and inversely proportional to the mass of the object.

Some of the terms in this may not be familiar to you, and they were certainly confusing to people in Newton's time. Consider "directly proportional to": The easiest way to explain it is to consider two quantities, A and B. If A is directly proportional to B, then as

Above: A tornado hitting a building is an example of a large force that acts over a very short period of time. Top left: Time-lapse images of an acceleration and deceleration test, which show the effects of rapid changes in acceleration on the human body.

A increases, B increases proportionately (for example, if you double A, then B also doubles). In the same way, when something (A) is inversely proportional to something else (B), it means that if A is doubled, B is cut in half.

When a billiard ball hits another billiard ball, it imparts a force to the second ball, causing it to accelerate.

IMPULSE

In the example of the boy on a swing, uniform force was applied. But there are many examples of forces that are not uniform. In fact, some are large forces that act over a very short period of time. Examples of this are a bat hitting a ball, a swiftly moving car hitting a resting car, or a tornado hitting a house. In each case the force is great, but it lasts only for a fraction of a second. Such forces are not uniform and indeed may vary considerably over the short time they are applied; so forces of this nature must be considered from a different point of view. We therefore introduce what is called *impulse*, which is defined as the product of the force and the time over which it acts. Since the time is so small, we can consider the force to be constant.

MOMENTUM

If an impulse involves a force and this force produces motion, we might expect that a given impulse would always produce the same amount of motion. This implies that the velocity of the object after the impulse is independent of its mass, which doesn't make sense. Our definition of impulse, however, does involve mass, since according to Newton's second law, force is the product of mass and acceleration. This means that impulse is therefore the product of mass and acceleration multiplied by the time over which the impulse lasts. But velocity is defined as acceleration multiplied by time, so we have a new quantity, namely the product of mass and velocity (mv), and it is called *momentum*. Impulse is therefore just the change in the momentum of an object.

When someone hits a ball with a bat, the bat imparts an impulse to the ball.

HITTING A BALL WITH A BAT

The concept of momentum can be applied to someone hitting a ball with a bat. The ball goes flying over the plate and the person swings at it. When the bat hits the ball, both objects are distorted to some degree, but only over a very short time. During this time the bat gives an impulse to the ball, and since the bat changes the direction and also the speed of the ball, the bat changes the ball's momentum.

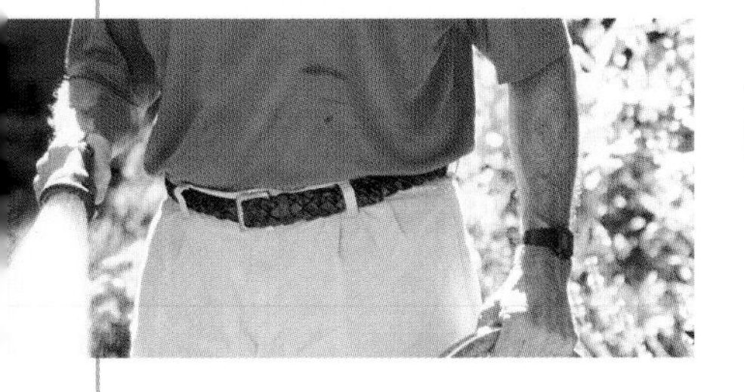

Momentum and Newton's Third Law

The concept of momentum is connected with another of Newton's laws. To see how, consider a force. A force always connects two bodies, one that is doing the forcing, and one that is receiving the force. Also, forces do not necessarily involve animate objects. In other words, there is not always a body applying the force. A block sitting on a floor also exerts a force, in a completely inanimate system. Indeed, if we look closely at the block sitting on the floor, we see that there are two forces and they are equal in magnitude, but opposite in direction. The block exerts a force on the floor because of its weight, but the floor is also exerting an upward force on the block. The same situation occurs when we push on something. We exert a force on an object and it then exerts an equal and opposite force back on our hands.

Newton recognized this and incorporated it into his third law of motion:

Whenever a body exerts a force on a second body, the second body exerts a force that is equal in magnitude but opposite in direction back on the first body.

These two forces are frequently referred to as action and reaction forces. We see examples of them every day. In holding a garden hose with water pouring out of it, you feel a backward force on your hand. This backward force is the reaction.

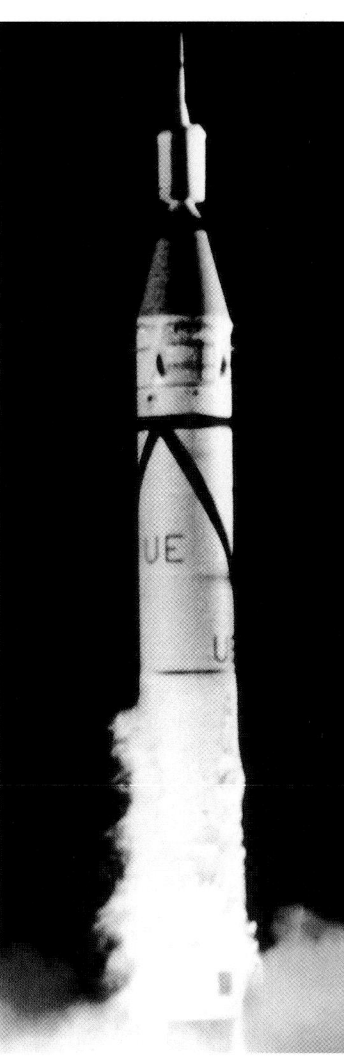

Above: A good example of Newton's third law: a rocket taking off to space. The gases shooting out of the back give the rocket a forward thrust. Top left: The action of water pouring from a hose causes the reaction of a backward force on the hand holding the hose.

To further understand the concept of action and reaction, consider rocket propulsion. Without a reaction, rockets could not be sent into space. They work because the gas that is shot out of the back of the rocket produces a forward thrust on the rocket. The burning fuel expands and exerts a force against the inner walls of the rocket, and the inner walls in turn exert an equal and opposite force against the gases. As a result of this force, the gases accelerate out the back of the rocket, so the rocket is forced to accelerate in the opposite direction.

CONSERVATION OF MOMENTUM

The principle of conservation of momentum can best be understood by considering an example. Assume that two cars collide. To keep things simple, assume they collide head-on, and that they both have the same momentum, but of course it will be in opposite directions. This does not mean they necessarily have the same speed, because their masses could be different. (We could, for example, have a slow-moving Cadillac

When two cars collide, as in this crash test, momentum must be conserved in the crash.

and a fast-moving Volkswagen.) When the two cars collide they will stop dead. What happens to their momentum at this point? Is it destroyed? No, it remains the same. Before the collision they had equal and opposite momenta, so their net momentum was zero. And after the collision it is still zero.

You could perform many such experiments, and you would always find that the momentum before equals the momentum after. This leads us to an important principle, called the law of conservation of momentum: *The total momentum of an isolated system of bodies remains constant.* In other words, in an interaction involving several bodies, the total momentum before the interaction will always be equal to the total momentum after, assuming there are no outside influences present.

How do we relate this to Newton's third law? Looking at the collision again, we see that when it occurred, each car imparted an impulse on the other, and the two impulses had to be equal and opposite to stop the cars. Furthermore, the time of the impact had to be the same for both cars, so the forces are equal and opposite. But as we saw, an impulse is just a sudden change in momentum, so the change in momentum of the first car has to be equal to that of the second car. Therefore the total momentum of the system is unchanged (the law of conservation of momentum), and the forces are equal and opposite (Newton's second law).

CENTER OF MASS

In addition to understanding what mass is, we must pinpoint the position of the mass. If the position is a point, there is no problem; but in most cases it is extended, and it may not be symmetric in shape (in a uniform symmetric body the mass can be considered to be at the geometric center). For problems of this type we have to determine the "center of mass"—in other words, the point where all the mass can be considered to be concentrated. This applies not only to asymmetric objects, but also to systems of particles (or objects).

How do we determine the center of mass? For something like a ruler, this is simply the balance point. For two-dimensional objects in general it can be determined by finding the point where the rotational force in one direction (clockwise) around that point is equal to the rotational force in the other direction (counterclockwise). This is generally straightforward in two dimensions, but in three dimensions it can become quite complicated.

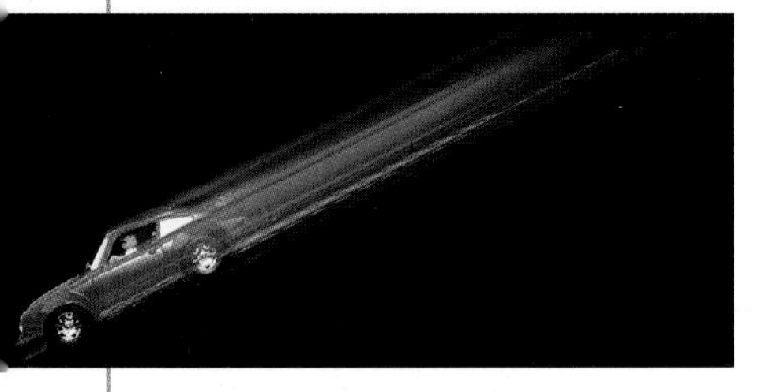

Gravity

Aristotle's model of the universe did not need gravity, since it contained the notion that all motions on Earth were "natural." A dropped stone moved toward the ground because this was its natural place. Aristotle ignored heavenly bodies because they were composed of a fifth essence and were unlike bodies on Earth. But Johannes Kepler (1571–1630), the German mathematician and astronomer, was seriously concerned with the objects in space and was puzzled about why they did not fall to the Earth. He realized that there had to be some sort of mysterious force operating out in space and thought that it might be the same force that he had observed on Earth.

Above: Three astronauts experience the weightlessness of space. They do not feel any gravity. With a small push they can float around the cabin. Anything in the cabin that is not tied down will also float. Top left: When Galileo investigated objects falling to Earth, he slowed them by sending them down an incline, like the ramp this toy car is on, rather than dropping them straight down from a height.

A leaf has a large area and is flat, so its fall to the Earth is slowed due to air resistance.

Galileo was also concerned with gravity, although his concern was mainly related to Earth. Why did objects appear to fall at the same rate (the exceptions being feathers, snow, and so on)? He wanted to measure the rate at which they fell but had a serious problem: He lacked a good timing device. The best he could do was to use his pulse. To get around the inaccuracy of his timing device, Galileo decided to slow the falling object. One way of doing this was to let it roll down an incline. The force (gravity) would still be acting on it, but it would now be measurable. Galileo soon noticed that no matter what the ball was made of, or how heavy it was, it always rolled down the ramp in the same amount of time. He also noticed that it sped up, or accelerated, at the same rate regardless of the mass.

Galileo also discovered something similar in relation to the pendulum. His pendulums consisted of weights attached to a long rope or string. He varied the weight on the end of the rope and found that it had no effect. The *period*, or time of swing, was the same regardless of the weight. The only thing that seemed to affect the period was the length of the rope: The longer it was, the longer the period.

Galileo was not able to measure the acceleration of gravity accurately, but today we are: It has an acceleration of 32 ft/sec^2 (10 m/sec^2) and is usually referred to as the letter *g*. This means that in one second a falling object will have a velocity of 32 ft/sec; in two seconds it will have a velocity of 64 ft/sec; and in three seconds a velocity of 96 ft/sec. Furthermore, using this formula, it is easy to find that if during the first second the object falls a distance of 16 feet, it will fall 64 feet in the first two seconds. The acceleration of gravity on Earth is, of course, characteristic of the Earth and is directly related to its mass. More massive planets such as Jupiter have higher accelerations of gravity, and lighter ones such as Mars have lower accelerations.

DIFFERENT PLANETS, DIFFERENT WEIGHTS

Your weight is determined by the acceleration of gravity of the Earth. On more massive planets such as Jupiter you would weigh more. Jupiter's gravity is 2.34 times that of Earth, so if you weigh 150 lbs on Earth, you would weigh 351 lbs on Jupiter. Similarly, the gravity of Mars is only 38 percent that of Earth, so on Mars you would weigh only 57 lbs. And since the Sun is 330,000 times more massive than Earth, you would weigh much more there, if you could weigh yourself before burning up.

Legend has it that Galileo performed experiments dropping objects from the top of the Leaning Tower in Pisa, Italy.

THE LEANING TOWER OF PISA

One of the most famous stories about Galileo is that he dropped balls of different weights from near the top of the Leaning Tower of Pisa and saw that they all fell to the ground in the same time. Aristotle had postulated that heavier objects fell faster than light ones, so Galileo's experiment proved him wrong. There is little evidence, however, that Galileo actually performed this experiment. It is known that a similar experiment was performed a few years earlier, in 1586, by Simon Stevin in Holland.

Frictional Forces

Everyone is familiar with friction in one form or another, as it plays an important role in our everyday lives. Friction is involved when we drive a car, walk, ski, skate, and even when we sit down on the couch and watch television. Friction has certain undesirable effects. For instance, the friction in the bearings and wheels, and the air resistance on a car, increases the work needed to keep the car going at a certain speed. So friction decreases the mileage we get. But without friction the car would not move. It is the friction between the tires and the road that gets the car going. Without friction we also would not be able to stand up. The friction between the soles of our shoes and the floor allows us to do this. And there are many other examples of the importance of friction—it keeps nails in boards and allows us to run machinery that uses drive belts.

In order to understand what causes friction, you have to look closely at the surfaces involved. The first property present on all surfaces is some degree of adhesion. Even if you cannot see them without the aid of a microscope, any irregularities on a surface contribute to adhesion and therefore friction. As the two surfaces slide past each other, the irregularities on one of the surfaces catch on those of the other surface. This causes each body to exert a force on the other. This force is referred to as a *frictional force.*

There are different types of frictional forces. Consider a box sitting on a surface, and assume that you have a scale with which you can measure the magnitude of a push or pull (a force) on the box. If you give the box a small push it may not move. The reason for this is that the frictional force between the box and floor is greater than the force you are exerting. If you push hard enough, though, the box will eventually move. You have overcome the frictional force, but you will have to continue pushing to keep the box moving because friction is still acting.

The two types of friction that are acting in this example are as follows: The frictional force between the floor and box before the box began to move is referred to as the *static frictional*

Above: When you push on a box, you have to overcome the frictional force between the box and the floor. Only when it is overcome will the box move. Top left: The friction between tires and road allows a car to move forward.

force. The force after the box began to move is called the *kinetic frictional force.*

PROPERTIES OF FRICTION

Through the example of the box being pushed along the floor, a number of important

In order to move, a locomotive wheel must experience friction. Without friction it will slip and the locomotive will remain stationary.

properties of friction can be explained. First of all, the *direction* of the frictional force is parallel to the surface. Second, the *magnitude* of the force is determined by the weight of the box, or, more exactly, the equal and opposite force acting upward, called the *normal force.* Third, the frictional force is independent of the area of contact, meaning that it does not depend on the area of the bottom of the box. Fourth, the force depends on the nature of the surfaces, or how rough or smooth they are—objects generally slide better when the surfaces are smooth. And finally, it does not matter how fast you slide the box, as long as you do not heat the surfaces; heat changes the properties of the surface and also the magnitude of the friction.

The friction between the road and a tire depends on specific conditions. Snow or ice decreases the friction dramatically.

STOPPING DISTANCE FOR A CAR

One occasion where friction is very important is when you want to stop a car suddenly, for example, to avoid an accident. When you want to stop a car suddenly, several types of friction are involved. One of the most important is the friction between the tires and the road, which can vary considerably depending on the condition of the tires (whether they are new or bald) and whether the road is dry, wet, or icy. The coefficient of friction for a good tire on a dry road is about .9, but when the road is wet it goes down to .7 or lower. If the road is icy, the coefficient may be as low as .1 or less. The stopping distance will therefore be longer if the road is wet and even longer if it is icy.

COEFFICIENT OF FRICTION

We can, in fact, define a coefficient called the coefficient of friction (designated as μ) that gives a measure of the "slipperiness" of the surface. The coefficient of friction is a number that is numerically equal to the force of the friction divided by the normal force between the object and the surface it is on. It depends on the two surfaces. Certain materials, such as ice, have a low coefficient of friction, while others, such as rubber, have a high coefficient. Coefficients of friction range between 0 and slightly over 1, with 0 specifying a surface with no friction and 1 specifying a very high friction. Actually there are two coefficients of friction—one corresponding to static friction (μ_s) and one to kinetic friction (μ_k). A few examples are as follows: steel on ice has $\mu_s = .02$ and $\mu_k = .01$, oak wood on oak wood has $\mu_s = .54$ and $\mu_k = .32$, and rubber on concrete (as in car tires) has $\mu_s = .90$ and $\mu_k = .70$.

Periodic Motion

The motion we have been talking about so far is unrestricted, meaning it has no bounds. But there is an important type of motion in physics that does have limits. This is back-and-forth, or vibratory, motion. When an object is continually in motion, and the same motion is repeated over and over, it is called *periodic motion*. Examples of this type of motion are a bob on the end of a spring moving back and forth or up and down, and a pendulum.

Periodic motion can be quite complicated. If you look closely you see that there are varying forces, varying accelerations, and varying velocities occurring throughout the motion. With all these variables, periodic motion is difficult to understand. It is best explained through an example. Consider a bob on the end of a spring that is hanging vertically. If you pull the bob down, the spring exerts an upward restoring force that pulls it back toward its equilibrium position. (In particular, the farther you pull it, the greater the restoring force.) After you let it go it begins moving back toward its equilibrium position, and as it moves, the restoring force decreases. Its acceleration also decreases since it is proportional to the force. Nevertheless, because of acceleration, the bob moves faster and faster as it approaches the equilibrium position. Finally, when it reaches this position, the restoring force has decreased to zero, and the acceleration is zero. But the bob now has its maximum velocity; in fact it has so much velocity there is no way it is going to stop (unless an external force is applied to it). It therefore rushes past the equilibrium position and continues moving upward. But as it passes this position, the force from the spring comes into play again. The force on the bob is directed toward the equilibrium position again, and, since this force causes an acceleration (actually, a negative acceleration), the bob slows down. As the bob continues to move upward, the restoring force continues to increase and the bob's velocity decreases. Finally, it stops. The bob is now at what is called its maximum displacement. At this point, the large restoring force in the spring starts pulling the bob back toward its equilibrium position, and again the force decreases as it approaches its equilibrium position.

The bob goes through this motion over and over, and if there are no external forces acting on it, it will continue indefinitely. In practice, however, there will always be some external forces present (mainly frictional forces), and they will gradually slow it down.

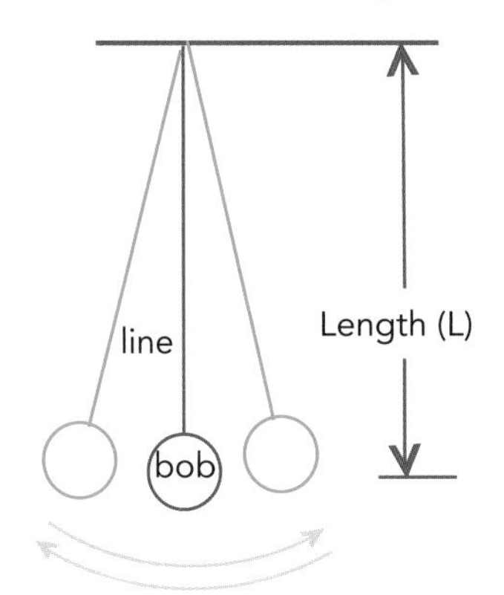

Above: A simple pendulum is an example of a simple harmonic system. Top left: A mass suspended on the end of a spring can be used to illustrate periodic motion.

SIMPLE HARMONIC MOTION

Any motion in which acceleration is proportional to displacement of an object and is always directed toward the object's equilibrium position, as in the case of the bob on a spring, is called *simple harmonic motion* (SHM). This is straight-line motion, in which the acceleration, velocity, and forces are continually changing. Very few systems exhibit perfect simple harmonic motion, but the motion of many systems is close enough to be considered simple harmonic. The bob on

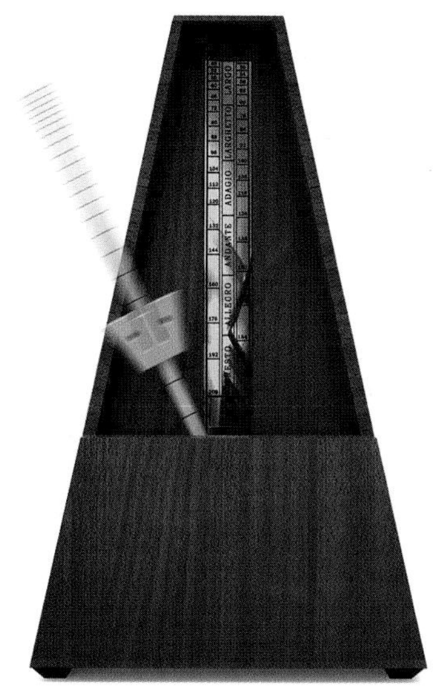

A metronome has a constant period of vibration. Musicians use metronomes to maintain a steady tempo.

the spring, a tight string that is pulled to one side and released, and a simple pendulum can all be considered simple harmonic systems. This is convenient because the mathematics is simpler in such systems.

PERIOD OF VIBRATION

In the example of the bob on a spring, the time it takes to go through the complete cycle—from the equilibrium position to one maximum displacement, back through equilibrium to the other maximum, and back to equilibrium—is called the *period of vibration.* It is usually designated by the letter *T*, and is usually given in seconds.

Closely associated with the period is *frequency.* This is usually designated as *f*, or sometimes $\boldsymbol{\nu}$. Frequency is the number of times that the system goes through a complete cycle in a unit of time, and it is represented by the unit vibrations per second (vibs/sec). Another important measurement related to periodic motion is the maximum displacement from equilibrium, which is referred to as the *amplitude.*

A carnival ride displaying pendulum-like motion—wherein a body swings back and forth under the influence of gravity after it is given an initial push.

Pendulums as timekeepers first came into use in the seventeenth century.

PENDULUMS AND CLOCKS

A good example of simple harmonic motion is a pendulum with a low amplitude. If amplitude is too high, a pendulum deviates considerably from simple harmonic motion. Galileo was one of the first to study the motion of a pendulum. Despite the fact that he noticed it had a very constant period, he never tried to use it as a clock. This was done for the first time by the Dutch scientist Christian Huygens, in 1673. His clock was quite ingenious in that it allowed for imperfection in the pendulum.

Circular Motion

Another type of motion that plays an important role in physics is circular motion—in other words, the motion of an object moving in a circle, or around an axis. In general this type of motion is more complicated than simple straight-line, or translational, motion. For instance, consider a rotating disk. Draw a line from the center of the disk to the edge, and label three points as A, B, and C, where A is relatively close to the center, B is farther out, and C is near the edge. When the disk starts rotating, the linear speed of point C is greater than that of point B, and similarly that of B is greater than that of A. This illustrates that in the case of rotational motion, speed depends on an object's distance from the center, or distance *r*. Therefore, distance *r* must be taken into account when defining rotational or *angular* speed.

ANGULAR SPEED AND ANGULAR ACCELERATION

Angular speed is defined as the number of revolutions or number of degrees rotated in a unit of time. The most common units are revolutions per minute (revs/min) or revolutions per second (revs/sec). Physicists also use a unit called the radian. By definition there are 2π radians (π is the circumference of a circle divided by its diameter; it has the value 3.1416) in one revolution (360 degrees), which means that one radian is $360/2\pi = 57.3$ degrees. Using radians, angular speed can be expressed using the unit radians per second (rads/sec). Angular speed is usually expressed as omega (ω), and again as in the case with linear speed, if we include the direction of the angular speed it becomes angular velocity.

Angular speed can vary, and when it does it becomes angular acceleration. Angular acceleration is defined as the rate of change of angular velocity. It is usually designated as alpha ($\propto$) and has units of revs/sec^2 or rads/sec^2.

Above: A rotating vinyl record undergoes circular motion. The farther out a point is from the center of the record, the greater its linear speed. Top left: A Ferris wheel moves in circular motion.

TORQUE

Simply put, *torque* is defined as rotational force—a force that causes circular motion. Since torque creates circular motion, its measurement depends on the distance from the axis of rotation to the point where the force is applied. Torque is therefore a force multiplied by distance (perpendicular distance from the line of force to the axis of rotation). It is usually designated as τ (tau). There are many examples of torque in everyday life.

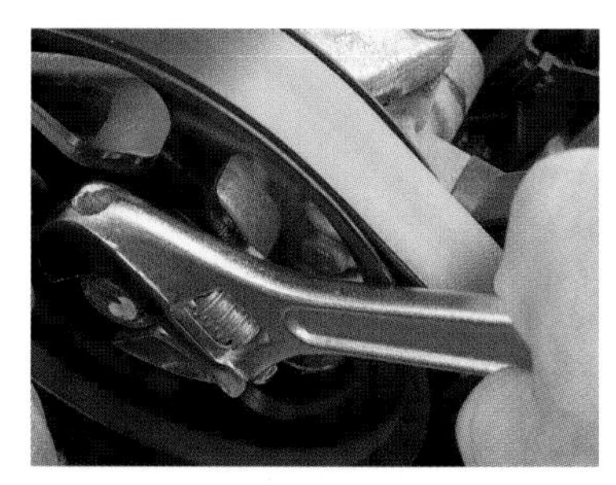

The best way to get a bolt loose is to apply torque to it using a wrench. Note that the farther your hand is from the bolt, the greater the torque.

You apply a torque when you turn a doorknob, twist the top off a jar, or use a wrench.

MOMENT OF INERTIA

Moment of inertia is a measure of a body's resistance to angular or rotational motion. It must be specified with respect to an axis of rotation.

The best way to understand moment of inertia is to use an analogy to linear motion. Newton's second law tells us that linear force is the product of mass and acceleration. In circular motion, these terms become torque and angular acceleration. Therefore, Newton's second law has an analogous form for angular motion, but there is an additional factor. While mass (call it m) played an important role in the linear case, with angular motion r must be included with mass, because the position of the mass is always important in angular motion. This is notated as mr. In short, masses at different distances from the axis have a different effect on the motion.

To determine a body's moment of inertia (I), we must add up all the (mr^2) subunits that make up the overall mass. In effect, there is an mr^2 for the mass of each point in the overall mass. An object's moment of inertia depends on its shape. A solid cylinder, for example, has a moment of inertia different from that of a hollow pipe, and a solid sphere has a moment of inertia different from that of a hollow sphere. In each case the mass is distributed differently, and there is a different amount of mass at a given distance (r). It's also important to note that the moment of inertia depends on the axis around which the object is rotating.

We now have what we need to modify Newton's second law for circular motion. Replace force by tau (τ), acceleration by alpha ($\propto$), and mass by I.

ANGULAR MOMENTUM

Earlier, we introduced momentum and defined it as mass multiplied by velocity (mv). In the case of angular motion we replace m with I, and v with $\propto$, which gives us angular momentum ($I\propto$). Also, as when linear momentum is conserved in any interaction between two or more objects, by analogy there is also conservation of angular momentum in that the total angular momentum of an isolated system of bodies remains constant. To illustrate this, assume you are standing on a turntable and have weights in each hand. Have someone push you to get the turntable spinning. When it gets going, you may be surprised to find that you have a strange control over its angular speed. If you extend your arms outward (with the weights in your hands) you will slow down; in other words, your spin will decrease. If you suddenly pull the weights in to your chest, the turntable will speed up. This occurs because angular momentum is conserved. Therefore, when r is decreased by pulling the weights inward, you speed up $\propto$, and when you increase r you slow down $\propto$. Figure skaters use this principle to speed up when they go into a spin.

A skater performing a spin. As he pulls his arms into his chest, his angular speed will increase.

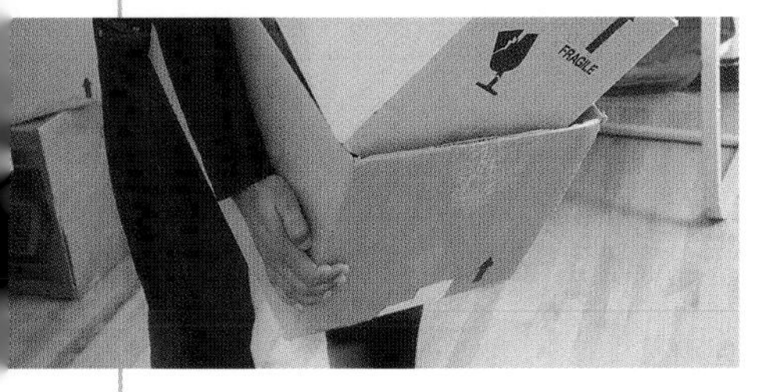

Work and Machines

The concept called *work* is used extensively in physics, but it may not be what you usually think of as work. Work in this case is defined as force (F) multiplied by distance (d), or Fd. Note that this is different from torque in that the distance is not necessarily perpendicular to the force; the distance in work is the displacement that takes place in the direction of the force. What is different in this definition of work is that if you lift a box (against the pull of gravity) to a counter, you are doing work; but if you merely carry it, you do no work, regardless of how tired your muscles get. If you slide it across the floor, however, you will be doing work against friction.

The above definition of work says nothing about how long it takes to perform the work. So time is important, and it is taken into consideration in what we call *power*. Power is the rate of doing work, so it is work divided by time.

The units of work again depend on the system we are using. It is foot-pounds in the British system, newton-meters in mks units, and dyne-centimeters in cgs units. Note that a newton-meter is a joule, and a dyne-cm is an erg, so the units of power are joules/sec in mks units and ergs/sec in cgs units.

If you are familiar with power in relation to cars, you know there is another power unit called the *horsepower*. Scottish engineer James Watt devised the unit of horsepower. He was interested in how much work a horse could do and found that a strong horse could lift 150 pounds through 220 feet in one minute. This is 33,000 ft-lbs/min, which is now called 1 horsepower.

Above: The unit of power called horsepower was arrived at by determining how much work a horse could do in a unit of time. Top left: When you lift a box, you are performing work.

It might seem a little ironic that the unit Watt devised is not named after him, but another power unit is. You are no doubt familiar with it in relation to your electric power bill. It is called the *watt*. By definition, a joule/sec is a watt, and 1 horsepower is 745.7 watts.

MACHINES

A machine can be defined as a device that makes work easier. Basically, it transforms a force from the point where it is applied to another point where it is used. There are, of course, many types of machines, including the inclined plane, the wheel and axle, the pulley, and various types of screws. Machines are, in fact, of two types, referred to as *simple* and *compound*. Simple machines perform work with one movement, while compound machines require more than one movement.

To illustrate the idea behind a machine, consider a lever. Suppose we are using it to raise a box. We place one end of a board under the box, and a few feet away we place a block called a fulcrum under the board. We then apply a downward force at the other end of the board, which raises the box.

Assume the distance from the box to the fulcrum is one-quarter of the distance from the fulcrum to where we apply the force. When we apply the force it will be applied to the bottom of the box, but we will not have to apply as much force as we normally would, because work is force multiplied by distance, and the distance the box is raised is only one-quarter of the distance that we pushed the board down. Thus we are expending extra distance to gain greater force. So the work done is the same, but it is easier for us because we apply less force. This principle applies to all machines; they make it easier to perform a particular task, but the total work is the same.

Furthermore, because of friction, the work done by the machine is less than the work done by us on the machine.

Simple gears. The force applied to the axle depends on the difference in size of the gears.

TYPES OF MACHINES

Machines play an important role in modern civilization. We could not, in fact, manage without them. They aid us in doing work that we might not otherwise be able to do. Some of the more important types of machines are as follows:

- Wheel and axle: A longer motion at the edge of the wheel is connected to a shorter, more powerful motion at the axle.
- Pulley: Combinations of pulleys permit heavy loads to be lifted by applying less force. The tradeoff is that you move the rope farther than the load moves.
- Screw: A screw converts rotary motion into forward motion.
- Gears: Gears are toothed wheels. Many different types exist. In a pair of gears the larger one will rotate more slowly than the smaller one, but with greater force. Other examples of machines are rack and pinions, wedges, and levers.

CHAPTER 2

PHYSICS AND THE SKY

Left: The night sky has always held great mystery for humankind. Over time, a number of visionaries made groundbreaking discoveries that brought us to the understanding of the universe that we have today. Top: The solar system, the definition of which has recently changed with regard to Pluto. Bottom: Modern observatories allow us to study the skies in a way that the ancients could never have imagined.

For thousands of years people looked out to the night sky and wondered what it all meant. Where did the Moon and the bright "wanderers" come from? What exactly were the Sun and the Earth and how big were they? How far away were the planets? Questions like these were the beginnings of human investigation into the heavens above.

There is evidence that early civilizations in China and Central and South America had an intense interest in the objects in the sky. By 750 BCE the Babylonians had noticed the phenomenon known as retrograde motion. In this motion, planets such as Mars suddenly stop their forward motion across the sky, move backward for a while, then turn around and begin moving forward again. This was confusing to the ancients, but they viewed it as the work of the gods and so it required no further explanation.

Over time, there were a number of visionaries who made dramatic advances toward a better understanding of the universe. Thanks to them, we now have a model of the solar system that has stood the test of time for hundreds of years.

Early Study of the Sky

The lives of early Babylonians centered on agriculture, so they studied the cycles of nature closely and eventually developed fairly accurate calendars. They carefully recorded the motions of the heavenly bodies and grouped the stars into constellations. The Egyptians did not make as many important observational advances as the Babylonians, but they did develop an excellent calendar. The flooding of the Nile was of particular importance to them as a source of irrigation. In order to better keep track of the cycles of this water source, they had to develop a good calendar. The Egyptians eventually developed a 12-month calendar with 30 days in each month and five days at the end of the year that were celebrated as a festival. Like the Babylonians, they considered the Sun, Moon, and other heavenly bodies to be directed by the gods.

EARLY GREEKS

The early Greeks were the first to create physical models of the solar system and the universe, and while some of

Above: Star fields of the Milky Way. When we look at the Milky Way at night we are seeing our galaxy. It has a diameter of 100,000 light years; we are situated 32,000 light years from the center. Top left: Pythagoras of Samos.

Above: Aristotle knew that the Earth was a sphere because he noticed that the shadow cast on the Moon during an eclipse was part of a circle.
Below: Fishermen drift along the banks of the Nile as twilight falls. As the sky darkens they will see a sky full of stars that are almost exactly the same as those seen by the ancients. The position of the stars has changed very little over the years.

the things they assumed might seem strange to us today, their predictions were amazingly accurate. One of the first Greeks of importance was Pythagoras of Samos, who was born c. 580 BCE. Most people are familiar with him because of his famous mathematical formula related to triangles, and he was indeed an excellent mathematician. But he also made important contributions to astronomy and formulated one of the first models of the solar system. At that time, eight heavenly bodies were known, including the Earth, Sun, Moon, and the planets out to Saturn. Pythagoras could not accept the fact that there were only eight—as a mathematician he viewed 10 as a perfect number and believed that therefore the number of bodies in the solar system should be 10. So Pythagoras invented two extra bodies to serve his purpose: a counter Earth and a central fire. Strangely, his model did not have the Earth at the center. The central fire was at the center but was not visible because it was blocked from view by the counter Earth.

Around the central fire were the Earth, Sun, Moon, counter Earth, and the known planets. Each was driven by an invisible celestial sphere, and (according to Pythagoras) each gave off a musical sound, which could be heard only by the gods. The stars were attached to the outermost sphere.

Aristotle (384–322 BCE) also put forward a model. In his model the Sun, Moon, planets, and stars were attached to invisible concentric celestial spheres that moved around the Earth. Aristotle was one of the first to assume a spherical Earth, and he based his assumption on solid evidence. He noticed that the shadow cast across the Moon during an eclipse was a section of a circle. Furthermore, he noticed that as an observer moved northward, the pole star appeared higher in the sky. Strangely, Aristotle also considered a heliocentric or Sun-centered solar system but rejected it for several reasons. If the Sun was at the center, the motions of the heavenly bodies across the sky had to be due to a rotation of the Earth, and he believed this was impossible. If the Earth was moving, he was sure we would be thrown off the surface and that the force would be so great that the Earth would eventually break up.

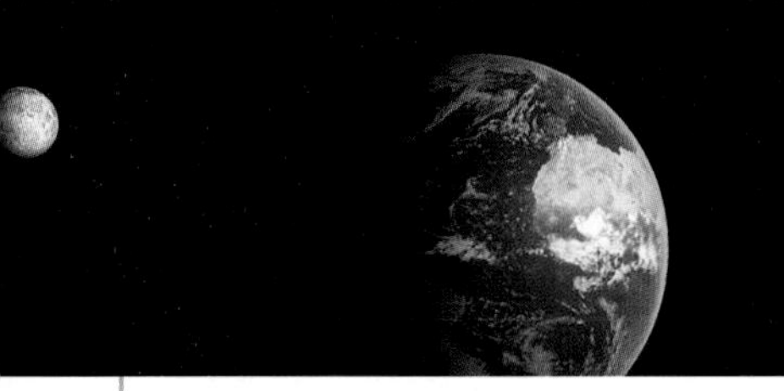

Modeling the Solar System

Of particular interest to the early Greeks was the size of the Earth, the Sun, and the Moon, as well as the distance of the Sun and the Moon from the Earth. The first person to arrive at an estimate for the size of the Earth was Eratosthenes of Alexandria, Egypt, who lived c. 276–194 BCE. He was visiting Syene in the south of Egypt on June 21 and noticed that at noon a vertical stick in the ground cast no shadow because the Sun was at its zenith, or directly overhead. He found that a similar stick on June 21 in Alexandria cast a shadow corresponding to an angle of 7 degrees from the zenith. If the Earth was flat, the shadows would have been the same at the two locations on the same date, but they were not. Eratosthenes therefore assumed that the Earth was a sphere. If so, the 7 degrees that he measured corresponded to 1/50 of a circle (7/360), and this meant that the circumference of the Earth was 50 times the distance between Syene and Alexandria. The distance between the two cities was given as 5,000 stadia, and therefore the Earth's circumference was 250,000 stadia. Unfortunately, we do not know how big a stadia was, but if it was 1/10 of a mile as some evidence suggests, Eratosthenes's estimate was quite accurate.

With this estimate and assuming that the Sun was 20 times farther away than the Moon, the Greeks could calculate the size of the Sun and Moon. Their estimates were close in the case of the Moon—.33 of the Earth's diameter, compared to the actual .27—but estimates for the Sun were 15 times smaller than its actual size.

THE PTOLEMAIC SYSTEM

One of the greatest Greek philosophers of this time was Ptolemy, who lived about 125 CE. He is best known for 15

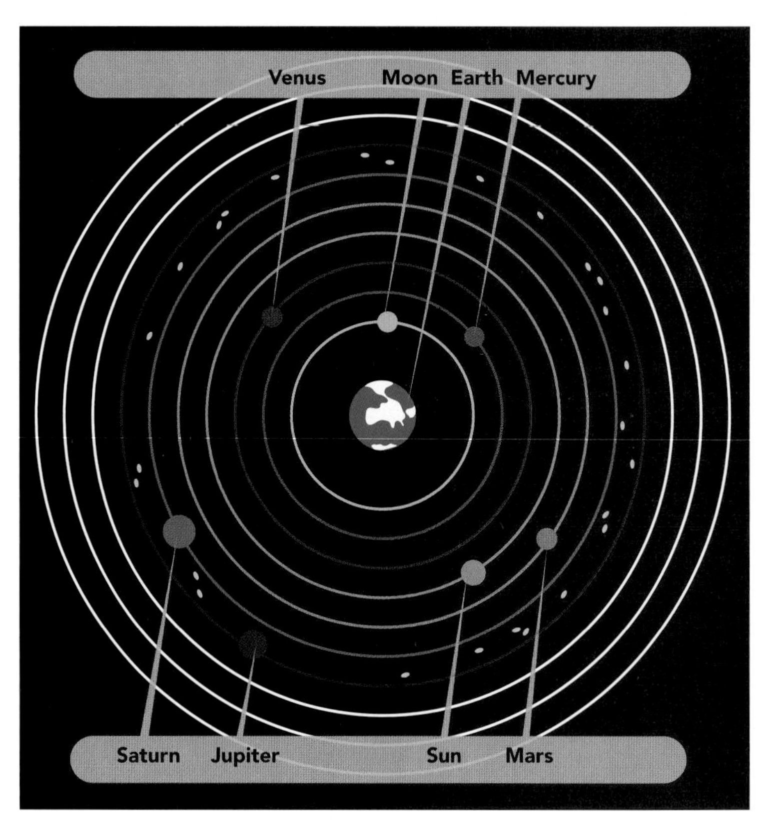

Above: Ptolemaic system with Earth at the center. Small epicycles (tiny circular orbits entered on the larger orbit) were used to explain retrograde motion. They are not shown in this representation. Top left: The ancients wondered about the size of our planet and the Moon.

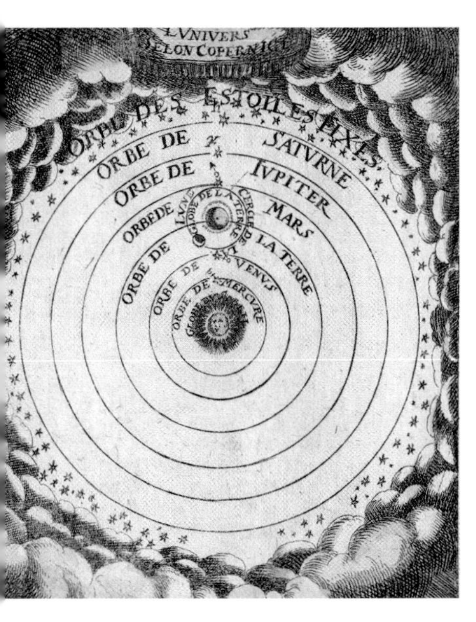

Copernicus's Sun-centered model of the solar system as depicted in his book De Revolutionibus. *The Copernican system was much simpler than Ptolemy's, but he was forced to retain small epicycles, which were eventually dropped from the system.*

volumes on astronomy and physics called the *Almagest.* His model of the solar system, which was described in the volumes, survived for almost a thousand years after it was put forward. It was an Earth-based system like the earlier systems, with the Sun, Moon, and planets attached to celestial spheres. Ptolemy's model, however, was considerably more accurate than earlier models; he could predict the future positions of all the known planets with considerable accuracy and used a concept called the epicycle to explain retrograde motion.

THE COPERNICAN SYSTEM

The Ptolemaic system stood until the early sixteenth century, which is a tremendous achieve-

Depiction of the Egyptian-Greek astronomer Claudius Ptolemy.

ment for any model. But it was not a simple model, and it was considered to be too complex by many people. Among them was Nicolaus Copernicus. Born in 1473 in Poland, Copernicus went on to train for a position in the church at the Universities of Krakow, Bologna, and Padua, but he was also interested in mathematics and astronomy. Copernicus studied the Ptolemaic system and eventually became convinced it was wrong. As an alternative he considered a Sun-centered (heliocentric) system, and he found that with it he could explain retrograde motion in a relatively simple way. After spending almost 20 years developing his model, Copernicus was reluctant to publish it because it went against the teachings of the church. Several people encouraged him, however, and finally when he was near death he began working on a book describing it. The book was titled *On the Revolution of Heavenly Spheres.*

NEW EXPLANATION OF RETROGRADE MOTION

As we saw, the early explanation of retrograde motion was based on the epicycle. With his heliocentric system, Copernicus was able to explain it in a simpler, more natural way. His explanation (which is illustrated for the case of Mars) is based on the idea that the Earth is traveling faster in its orbit than Mars. In this diagram, the Earth's orbit around the Sun is indicated in blue, and Mars's orbit is in red. At first, Mars appears to be moving eastward in the sky (right to left in the diagram). But since Earth is moving faster, it overtakes and passes Mars, causing Mars to appear to slow down, stop, and change direction in what is called a retrograde motion. Then, as the Earth moves ahead of Mars again, Mars resumes its eastward motion.

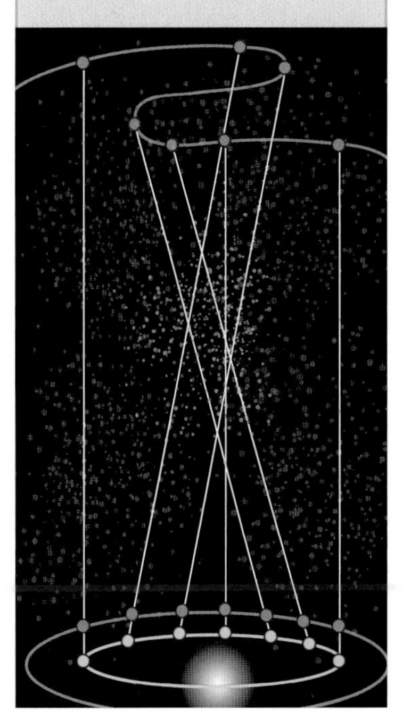

Kepler's Laws of Planetary Motion

Copernicus's model of the solar system was around for years before people began to take it seriously. The main problem was that there was no proof it was better than the Ptolemaic system. But this was soon to change, and the change was initiated by two men. The first was Tycho Brahe (1546–1601) of Denmark. Tycho (as he is called) studied law, but his real love was astronomy, and when he observed a nova, or new star, in 1572 he wrote a book about it, *De Nova et Nullius Aevi Memoria Prius Visa Stella* (*On the New and Never Previously Seen Star*). The book caught the attention of Frederick II, king of Denmark, who was so impressed he granted Tycho an island so that he could set up an observatory. Tycho called it Uraniborg, and it eventually became the best-equipped observatory in the world. Although the telescope had not yet been invented, Tycho had the world's best sextants to measure the positions of the planets and stars, and he had extremely accurate clocks. As a result, over many years he accumulated a tremendous amount of astronomical data on the planets, the Sun, Moon, and the stars. He did not, however, have the mathematical background to use the data to formulate a model of the solar system. Fortunately, he was eventually joined in his research by someone who did: Johannes Kepler.

Born in Weil, Germany, in 1571, Kepler made important contributions to astronomy, and his writings caught Tycho's eye. After Tycho left Uraniborg and resettled in Prague, he invited Kepler to work for him. Kepler knew of the data that Tycho had accumulated over the years and was eager to see it, but to his dismay the protective Tycho gave him little access to it. Within a year, however, Tycho died, and his data, along with his position as imperial astronomer of the Holy Roman Empire, was inherited by Kepler.

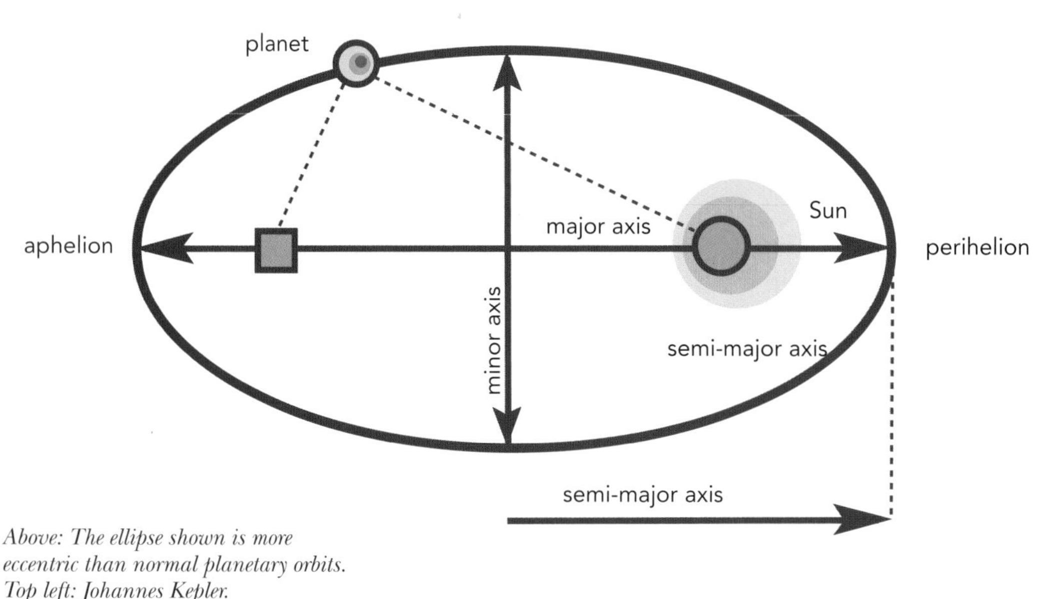

Above: The ellipse shown is more eccentric than normal planetary orbits. Top left: Johannes Kepler.

KEPLER, MARS, AND THE THREE LAWS

Kepler was delighted. He now had what he needed to investigate the ideas he had about the solar system. Using Tycho's data, he began with the planet Mars. For years he tried to fit its orbit to a circle, but there were always discrepancies. Something was wrong. Eventually he tried an oval-like orbit called an ellipse, and it worked. Mars was not tracing out a circle; it was moving in an ellipse.

Kepler came to understand that all the planets were moving around the Sun in elliptical orbits. The ellipses, however, were very close to circles—so much so, they are almost indistinguishable from circles. Nevertheless, to the accuracy that Kepler was working, their elliptical shape was important. He published his results in 1609 in a book titled *New Astronomy*. In all, Kepler arrived at three important results that he formulated as laws. They are now referred to as Kepler's laws.

Kepler's first law states: *The orbit of each planet is an ellipse with the Sun at one foci.* Foci are the two points symmetrically located on the major axis of an ellipse at either side of the center. Kepler's second law is: *The radius vector of each planet sweeps out equal areas in equal times.* The radius vector is the line (or vector) from the Sun to the planet. The consequence of this law is that the planet is not orbiting the Sun with a uniform speed. Rather, it moves faster when it is close to the Sun and slower when farther away. The point in a planet's orbit where it is closest to the Sun is called the perihelion. The planet's speed is fastest at this point. The most distant point is called the aphelion, where the speed is slowest.

And Kepler's third law states: *The squares of the orbital periods of the planets are proportional to the cubes of their mean (average) distances from the Sun.* If we designate the period of the planet by T and its mean distance by a, this tells us that the period squared ($T \times T$) divided by the mean distance to the planet cubed ($a \times a \times a$) is a constant for all orbits. In formula form this is T^2/a^3.

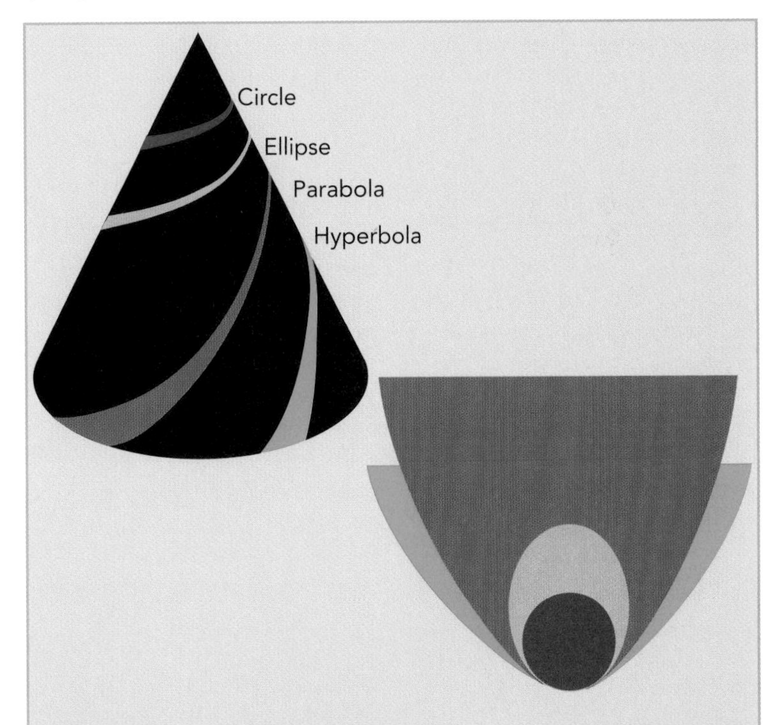

The conics, shown as slices through a cone.

CONICS AND ECCENTRICITIES

Kepler's three laws are particularly helpful in determining orbits in space, which are one of four possible shapes: a circle, an ellipse, a parabola, or a hyperbola. The best way to visualize these curves is to imagine a cone being sliced. If you cut it parallel to the base you get a circle; if you cut it at an angle you get an ellipse. Similarly, if you cut it so the cut goes through the base at an angle you get a parabola, and if you cut it perpendicular to the base you get a hyperbola.

The eccentricity of a given curve (designated as *e*) gives a measure of its elongation or shape. Circles have an eccentricity of zero; the eccentricity of an ellipse can vary from 0 to 1; the eccentricity of a parabola is 1; and a hyperbola has an eccentricity greater than 1.

Newton and Universal Gravitation

Newton not only revolutionized our understanding of motion on Earth, but he made a discovery that revolutionized our views of motion in space as well. Kepler was sure there was a force in space that acted on the planets, but Newton demonstrated how it worked and was able to predict its strength.

THE LAW OF GRAVITATION

Newton began working on the problem of a mysterious force in space when he returned to Woolsthorpe during the plague. He was convinced that there was a force between the planets, and he now had his three laws of motion to help him work out its nature. Newton soon realized that this force had to be pulling the planets toward the Sun, and it also had to be pulling the Moon toward the Earth. Combining his second law of motion with Kepler's third law, Newton was able to figure out how the force weakened with distance.

According to legend, Newton was sitting in his garden one day thinking about the problem when an apple from a nearby tree fell to the ground. He thought about it. The apple accelerated as it fell, so there had to be a force pulling it toward the ground. Was this the same force that pulled the Moon to the Earth, and the Earth to the Sun? An assumption such as this was a bold leap, but Newton took the leap and incorporated this new idea into a new law of nature, the law of gravity:

Every particle of matter in the universe attracts every other particle of matter with a force that is proportional to the product of their masses and inversely proportional to the square of the distance between them.

The mathematical formula for calculating this is: Force = G $(m_1 m_2)/r^2$, where G is a universal constant of gravity, m_1 and m_2 are the masses of any two bodies, and r^2 is the square of the distance between them. It is this inverse square law that causes the planets to move in ellipses; in other words, the inverse square law leads to elliptical motion.

Above: Our solar system as seen from space. This illustration shows the elliptical orbits of the planets. Top left: Earth's Moon.

CALCULATION OF THE MOON'S PERIOD

Newton decided to test his new law on the Moon. The period of the Moon was known to be slightly over 27 days, and Newton was sure he could calculate this from his law. He knew that the Moon was 60 times farther from the center of the Earth than it was from the surface of the Earth, and if the force fell off as the square of the distance, the gravitational force at the Moon would be $1/60^2$ (1/3,600) times the gravitational force at the surface of the Earth. Since mass would be the same at both locations, and the acceleration of gravity is 32 ft/sec^2 on Earth, it would be .00028 ft/sec^2 at the Moon. This would be numerically equal to its acceleration, and from its acceleration Newton was able to calculate the period of the Moon. It was close to the observed 27 days, but he was disappointed that it was not completely accurate. The reason for the inaccuracy was that the distances and the acceleration of gravity were not known to a high degree of accuracy at the time. With their accurate values he would have arrived at an exact value for the period.

Newton's theory of gravitation was a tremendous breakthrough in our understanding of the universe. Two new planets were eventually discovered as a result of predictions based on it. The orbits of the "mysteriously disappearing" comets could also be calculated, explaining why they disappeared. It was truly a milestone in the history of science.

Galileo at his telescope. Galileo made many important astronomical discoveries, including the four largest moons of Jupiter, the mountains on the Moon, and the phases of Venus.

GALILEO AND THE COPERNICAN SYSTEM

Galileo also played an important role in the new revolution. He had speculated on gravity but did not understand it well. Nevertheless, his observations were critical in overthrowing the Ptolemaic system and ushering in the acceptance of the Copernican system. Galileo did not invent the telescope, but he improved on it significantly and was the first to turn it toward the heavens. He noticed the phases of Venus and the four tiny moons orbiting Jupiter, two things that convinced him that the Copernican system was correct and the Ptolemaic system was not. Like Copernicus, he was cautious in sharing his views. Eventually, however, he published a book titled *Dialogue on the Two Chief Systems of the World*, in which he presented his "proof" of the Copernican system. This angered the pope, who ordered cancellation of the publication. Galileo was summoned before the Inquisition and was forced to sign a document saying he denied his views on the Copernican system. He spent the rest of his days under house arrest.

CHAPTER 3

WAVES

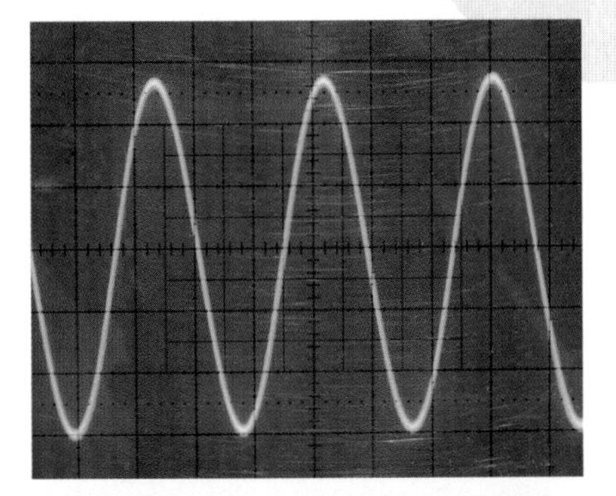

Left: Water waves are only one of the types of waves that exist in the world around us. The properties of these waves are shared by all other types. Top: A sine wave on an oscilloscope. Bottom: Waves are sometimes affected by objects passing through them, such as a fishing lure on a pond.

Waves are all around us: sound waves, radio waves, microwaves, water waves, earthquake waves, and even the gesture we make when we say good-bye to someone. Waves are so familiar that people generally do not give them much thought, but they are an important phenomenon in physics.

One of the best places to study a wave is on a calm lake or pond. If you throw a stone into the water, you see waves expand out from the point where the stone hits the surface. As the stone enters the pond, it pushes the water down; and since water is incompressible, it has to be pushed somewhere. In this case the water is pushed outward, and this outward surge produces a raised ring. This ring, which is above the normal surface of the pond, is pulled down by gravity, producing another ring farther out. Each ring, in turn, creates another, which creates the ripples that we observe on the surface of the water.

All waves have the same general properties, so this water wave is representative of all waves.

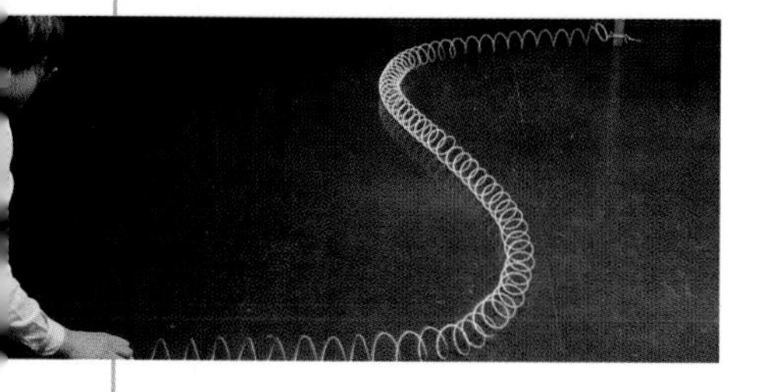

Wave Types

There are two types of wave: transverse and longitudinal. The former is characterized by motion that is perpendicular to the direction in which the wave is traveling. A longitudinal wave moves parallel to the direction of the motion of the waves.

TRANSVERSE WAVES

One of the best ways to understand a transverse wave is to create one and examine it in detail. We can do this using a rope. Begin by tying the rope to a doorknob or other projection and pulling it tight. With a quick jerk in an up-and-down direction, you will generate a pulse that will move down the rope to its end. In fact, if you move your hand up and down quickly several times, you will send several pulses down the rope one after the other. What we would like to do is send a continuous sequence of them, and we can do this by using a rod or probe that vibrates with simple harmonic motion. By attaching this harmonic motion machine to the end of the rope and setting it going, we will get equally spaced pulses moving down the rope. This is a wave, and it has the same properties as the water waves we have examined. Looking at it closely, we observe a series of equally spaced *crests* and *troughs*. The top section of the wave is referred to as a crest, the bottom section as a trough.

There are also points where the rope is not displaced from its equilibrium position. These points, called *nodes*, are either descending or ascending. A node on the downsloping part of the waves is called a descending node; one on the upsloping section is called an ascending node.

The distance from the equilibrium position to the top of the crest is referred to as the *amplitude*. Also, if you were standing at some point along the rope, a certain number of crests would pass you every second, so the wave obviously has a certain velocity. This velocity, it turns out, depends on the properties of the rope.

Above: This diagram of a transverse wave shows several of the properties of a wave: amplitude, wavelength, a crest, and a trough. Top left: One way to illustrate a transverse wave is to connect one end of a flexible spring to a chair while holding the other end and shaking it to create the wave motion.

LONGITUDINAL WAVES

To see how longitudinal waves work, we can use the example of a coiled spring (such as a Slinky) attached to a doorknob or wall and stretched out so that it is parallel to the floor. If we hit the end, a pulse moves down the coil with a speed that depends on the physical

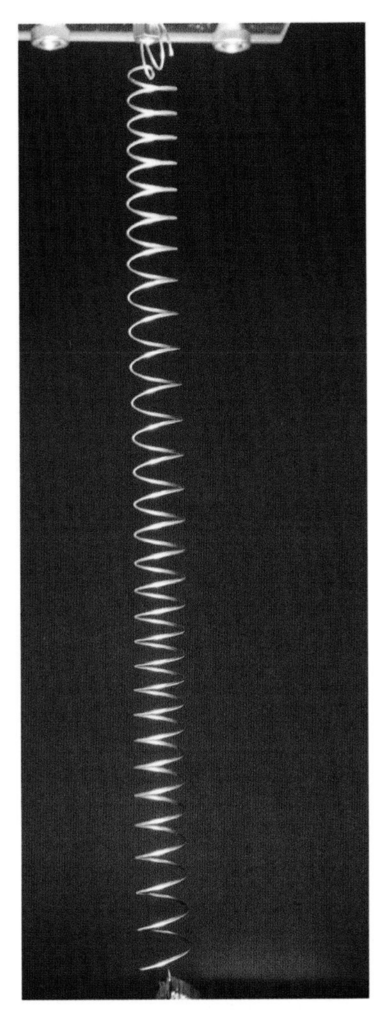

Longitudinal waves are easy to illustrate using springs like this one. Bang on one end and a longitudinal wave will travel along it.

The breakers that we observe on the beach are the result of a number of things that occur as waves approach the shore.

OCEAN WAVES

A good place to see large waves is on the ocean. Again, they look like they are moving across the surface, but the individual particles of water move only a small distance horizontally from their equilibrium position. Indeed, if you put a cork on the water it will move up and down with only a slight forward and backward motion.

When they hit shallow areas several things occur that cause the breakers we observe at the beach. A wave grows dramatically when the bottom of the wave hits the sand beneath the water. When the bottom of a wave starts to hit the sand, the wave slows down and other waves crowd in behind it, causing the waves to squeeze together. At the same time, the back of the wave is now traveling faster than the front, which causes it to rise to a peak. This peak then begins to curl and spills over, or breaks. A wave will normally break when the ratio of its height to the water depth is 3:4.

properties of the coil. And, as in the case of the rope, we can attach a simple harmonic motion vibrator to the end and get a series of equally spaced pulses moving down the coil. This motion is a longitudinal wave, and in many ways it is not as easy to visualize as a transverse wave. It has all the same properties as the transverse wave, but different terminology is used. In place of the crest we have a *compression*, where the sections of the spring are closer together than normal; and in place of the trough is a *rarefaction*, where the sections are farther apart than normal. We also have nodes, where the spring is at its equilibrium position. Furthermore, the distance from a node to the maximum of a compression or rarefaction is the amplitude.

A good example of a longitudinal wave is a sound wave. In this case, a vibrator, such as our vocal cords, sets up a wave that moves through air. We can create the same type of waves in a tube of air with a piston at one end. A sudden movement of the piston will send a wave through the air down the tube. Longitudinal waves are able to travel through gases, liquids, and solids.

Frequency and Wavelength

Waves have a number of other properties that we have not yet described, which apply to both transverse and longitudinal waves. The distance between one crest and the next (or between one trough and the next, or any two equivalent points) is known as *wavelength.* In the case of a longitudinal wave, it is the distance between one condensation, or rarefaction, and the next. Wavelength is usually designated as λ.

Closely associated with wavelength is *frequency.* If we stand at some point along the wave, a certain number of crests will pass us per unit of time. This is the wave's frequency, designated as f. The units for frequency are number per second. Period, or T, is the reciprocal of frequency, and it has units of time. It is the time that it takes for one wavelength to pass. Finally, we know that the wave is traveling at a certain velocity (v). The relationship between f, λ, and v is represented by the formula $v = \lambda f$, or velocity equals wavelength multiplied by frequency.

SOUND

Sound requires a medium to transmit it. In most cases this medium is air. A sound wave is created when something is struck and vibrates; this causes the air to vibrate, creating a longitudinal wave. The most familiar type of longitudinal wave is sound, which can be defined as a disturbance or vibration of air molecules that is detected by the ear. The ear can detect frequencies from about 20 vibrations per second (vib/sec) up to approximately 15,000 vib/sec. The upper limit depends to a large degree on one's age; it decreases as we get older.

Like any wave, sound has a certain speed. The speed of sound through air (at 68°F, or 20°C) is 1,130 ft/sec (344 m/sec). This speed varies slightly

Above: When a train sounds its whistle you hear a distinct change in its pitch as it approaches or travels away from you. The pitch increases when it approaches and decreases when it recedes. Top left: Sound is a longitudinal wave that is detected by the ear.

with temperature and pressure and also varies when the waves are traveling in media other than air. In water, for example, the speed of sound is 4,757 ft/sec (1,450 m/sec), and in iron it is an incredible 16,732 ft/sec (5,100 m/sec), which is why, long before you can see it, you can hear a train coming by putting your head to the rails.

Speeds greater than the speed of sound are referred to as *supersonic.* Airplanes are now capable of speeds several times that of sound. When we refer to such speeds we usually use the designation Mach, named for the Austrian physicist Ernst Mach, who first investigated them. The speed of sound is referred to as Mach 1, double its speed is Mach 2, and so on. Airplanes now travel at speeds greater than Mach 3, and astronauts in space travel at speeds greater than Mach 25.

A sound wave can also bend as it travels through the atmosphere or any other medium. This bending is referred to as *refraction.* An interesting example of this sometimes occurs in the evening when the upper layers of the atmosphere are warmer than the lower layers. When sound hits these warmer layers it speeds up and is deflected downward. This is why in the evening you can sometimes hear the voices of people in a boat out on a lake very clearly, even though they are a long distance away. This natural amplification over a cool body of water is one of the few natural examples of sound refraction.

Christian Doppler (1803–53), discoverer of the Doppler effect.

DOPPLER EFFECT

Another important phenomenon that we encounter in relation to sound is called the Doppler effect, named for the Austrian physicist Christian Doppler, who first studied and explained it in 1842. If you are standing at the edge of the road and a car with its horn blaring passes you, there is a sudden change in the sound's pitch as it passes. This change in pitch is the Doppler effect. To see why this occurs, consider the waves that are being emitted by the horn. If the car was at rest relative to you, the wavelength of this sound would be the same in all directions. But since the source of the sound is moving, it is "catching up" slightly with the waves in the direction it is traveling. This causes the wavelength to decrease slightly in this direction (and increase slightly in the opposite direction). Because of this, as the car approaches, its horn has a slightly higher pitch than it would have had at rest. Furthermore, as it passes, the pitch decreases to a lower pitch than normal. What you are hearing in a change of pitch is a change in frequency—longer wavelengths are perceived as lower in pitch than shorter ones. The same effect occurs if you are moving and the source of sound is stationary or if the transmitting media (the air) is moving, as when it is windy.

Frequency and Music

How does the sound of music differ from the numerous other sounds we hear? One major distinction is that the frequency of sound emitted by musical instruments does not change suddenly. Other types of sounds, on the other hand, change frequency quickly and often.

MUSICAL TONES

A musical note has a certain *pitch* associated with it. For our purposes, pitch is the same as frequency—the higher the frequency, the higher the pitch; the lower the frequency, the lower the pitch. Frequency plays an important role in music, as each different note has its own frequency, and it is this variation in frequency that creates a melody.

Closely associated with frequency or pitch is *quality* or *timbre* of a musical sound. Consider a note such as middle C on the piano. We know it has a frequency of 256 vib/sec, but if you sound this pitch on different instruments, such as a violin, a trumpet, or a clarinet, you notice that it sounds different on each. The pitch on all three instruments still has a frequency of 256 vib/sec, but the so-called envelope that makes up the sound in each case is shaped differently. To see why, we have to look at overtones, which are musical tones that help to determine the overall quality of the sound.

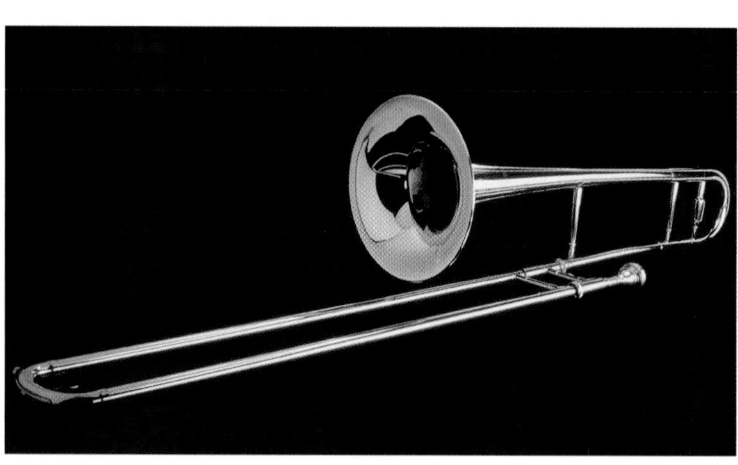

Above: When a note is sounded on an instrument such as a trombone, it sounds different from a note sounded on a piano or violin. The reason is the variety of overtones that accompany the note. Top left: The amplified sound of an electric guitar illustrates the property of loudness, or intensity, of sound.

OVERTONES

When we sound the note C on the piano, a hammer hits a string (actually, in most cases, a set of three strings) of a certain length and causes it to vibrate. The vibrational modes that are set up can be seen by observing the string. When the string vibrates with one loop, or point where the wave is the widest, and nodes at each end we get what is called the "fundamental" tone. But it is also possible for this wire to vibrate in other ways. It can, for example, vibrate with two loops and a node at the center (and at the ends); this is referred to as the first overtone. Similarly, it can vibrate with three loops; this is the second overtone. In practice, these overtones, or harmonics, as they are also called, are all vibrating at once. The fundamental, or first harmonic, is dominant in that it is the loudest; but the second, third, and subsequent harmonics are also present in the string. Because of this, middle C is actually a superposition of many harmonics, and they determine its overall shape, or envelope.

This also occurs in the case of the violin, the clarinet, and the trumpet, but the overtones or higher harmonics are different in each case, so the envelope is different.

Another important property of a musical tone is loudness, or intensity, which is the rate at which the sound energy flows through a unit area. In practice, the loudness of a sound depends on both intensity and frequency. The reason for this is that the ear has a different sensitivity to different frequencies. The most sensitive range is between 2,000 and 4,000 vib/sec; very low and very high frequencies are generally harder to hear.

A tuning fork. This tool is used to sound the note A, in order for orchestra members to tune their instruments.

The decibel rating for the takeoff of a large jet is particularly high, usually measuring approximately 140.

DECIBEL SCALE

The range of sounds that impinge on our ears each day varies considerably. The unit that is usually used to measure them is called the bel, in honor of Alexander Graham Bell, who invented the telephone. If one sound has 10 times as much power as a second sound, the difference in intensity of the two sounds is said to be 1 bel. If the first sound is 100 times as great, it is 2 bels, and so on. (Such a scale is said to be logarithmic. A logarithmic scale is a scale that uses the logarithm of a physical quantity instead of the quantity itself.)

Since a bel is a particularly large unit, it is sometimes more convenient to use the decibel (db), which is one-tenth of 1 bel. On this scale, as we go up one unit (1 db) we increase the intensity of the sound by 26 percent. The following table gives some examples of decibel ratings for various sounds.

Source	Decibels
Smallest change ear can detect	1
Average living room	40
Ordinary conversation	70
Street traffic in large city	80
Factory	100
Threshold of pain	120
Takeoff of jet	140

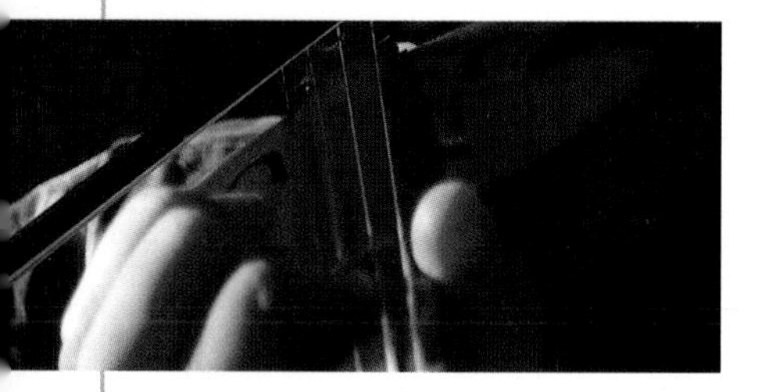

Musical Scales and Instruments

Music consists of a succession of tones that are pleasing to the ear. We refer to such a succession as a melody, but to produce a melody we need a musical scale. In this scale only certain frequencies occur.

The Greek philosopher Pythagoras was the first to show that certain frequencies or notes sound pleasant when played together. He was unable to measure frequency, but he did experiment with musical sounds produced when strings of certain lengths were plucked. He found that two strings in the ratio 1:2 produced a pleasant sound when plucked together. He also noticed that other ratios, such as 2:3, 3:4, 4:5, and so on, also produced pleasant sounds.

The scale we use today, called the diatonic scale, is based on Pythagoras's discoveries. It has eight notes, in which the ratios 3:2, 4:5, 5:3, 5:4, 9:8, and 15:8 play an important role. Using middle C (256 vib/sec) again as the example, when the above ratios are applied to the properties of middle C, the other notes of the scale result. Commonly referred to as the letters D, E, F, G, A, B, C, their frequencies are 288, 320, 341, 384, 427, 480, and 512 vib/sec, respectively.

INSTRUMENTS

Musical sounds can be produced in many ways. In a piano, a hammer strikes wire strings that are under considerable tension.

In a violin, strings also vibrate, but this time we change their frequency by changing their length with our fingers, and we create the sound with a bow or by plucking the strings. Musical sounds can also be produced using tubes, as in wind instruments, where sound waves enter tubes that can then be altered to produce the resulting sound.

The slide of a trombone lengthens or shortens the tube, while a flute player blocks holes in the instrument to alter the sound. Earlier we saw how harmonics could be set up on strings. In the same way they can be set up using tubes. There are three configurations of tubes to consider—a tube with both ends open, one with an open end and a closed one, and a tube with two closed ends.

We will look at the first two cases. Note that there will always be a loop at an open end, and that nodes always occur within the tube. For two open ends (figure A), three cases are shown. In the top

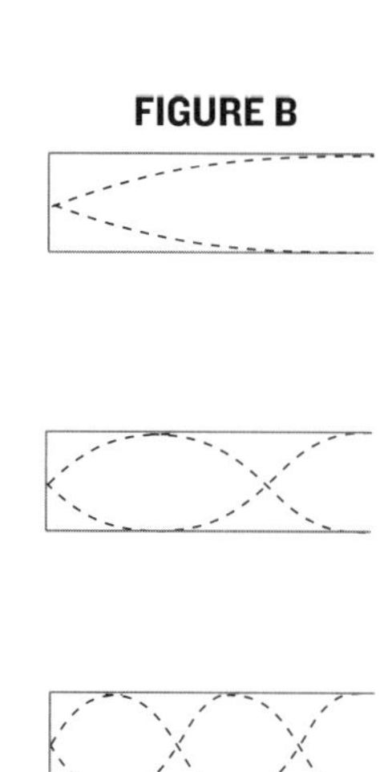

Above: These figures illustrate how harmonics work in a tube, as they would in a wind or brass instrument. Top left: Violinists change pitch by changing the length of the strings.

one we see that the length of the tube is equal to half a wavelength, in the middle one the length is equal to the wavelength, and in the bottom one the length is equal to two wavelengths. For an open end and a closed end (figure B), again three cases are shown. In the top one the length of the tube is one-quarter of a wavelength, in the middle one it is three-quarters of a wavelength, and in the bottom one it is five-fourths of a wavelength.

ACOUSTICS

It is important that a concert hall have good acoustics. In other words, the music has to sound vibrant and full of life, and one of the things that accomplishes this is the number of reflections the sound makes from the walls before it dies away. These reflections, collectively referred to as reverberation, are affected by the texture of the surfaces off which the sound is reflected. Hard surfaces such as marble or plaster reflect well and give a high reverberation. Porous surfaces such as those of fiberboard or drapes have a high absorption coefficient and give low reverberation. Excess reverberation is undesirable, but if there is no reverberation at all, a concert would sound muffled, or dead; so a certain level of reverberation is necessary to produce a pleasing sound. Creating a hall with great acoustics therefore requires the measurement of reverberation.

Music halls are designed to have good acoustics. The type of materials used for the walls is of particular importance.

WALLACE SABINE'S ACOUSTICS EXPERIMENT

One of the first physicists to consider reverberation in concert halls was Wallace Sabine of Harvard University. In 1895 the university opened a new auditorium, but it was soon discovered to have poor acoustics. Sabine was assigned to find out what the problem was. He began a long series of experiments and built up a large amount of knowledge on acoustics. In particular, he studied the role of reverberation time (time to inaudibility, or zero sound) and determined that it was particularly important. Because of his experiments, he was called in as a consultant when the nearby Boston Symphony Hall was being constructed. He now had a basic technique for determining the best reverberation time, and he applied it to the new hall, sure that he would make it acoustically sound. To his disappointment, when the hall was opened in 1900 he was severely criticized for his design. As it turned out, he had designed a hall with excellent acoustics but had failed to take the audience into consideration. The presence of an audience changed the reverberation time of 2.3 seconds to 1.8 seconds, which had a serious effect on the sound. Sabine was devastated and never designed another hall. Later, however, it was found that his techniques were indeed a breakthrough in concert hall design, and they are still used today.

Wave Interference

Waves can interfere in two different ways: constructively and destructively. When waves interact they produce new waves. This interaction is referred to as the superposition of waves, and it occurs when two or more waves come together in the same medium. The result of superposition is a summation of the displacement of the waves. The best way to see this is to consider two transverse waves. For simplicity, assume the two waves are of the same frequency and have the same amplitude. If they are in phase, with the crests lined up, the two waves will come together and produce a wave that has twice the amplitude of the two original waves. This phenomenon is called constructive interference.

However, if the two waves come together half a wavelength out of phase, the crest above the axis will cancel the trough below it, and a lack of motion results. This is referred to as destructive interference. In most instances, when two waves come together their frequencies and amplitudes are not exactly the same, so something in between the two cases results.

Above: Waves on water, like other types of waves, interfere with one another, as seen here. Top left: The wake of a boat disrupts normal wave patterns.

A case of particular interest is when two waves of slightly different frequencies start off at the same time. At first the two are generally in phase, but they soon fall out of phase and begin canceling each other out, then

A vibrating string showing nodes and loops. Note that there is a node at either end and in the center.

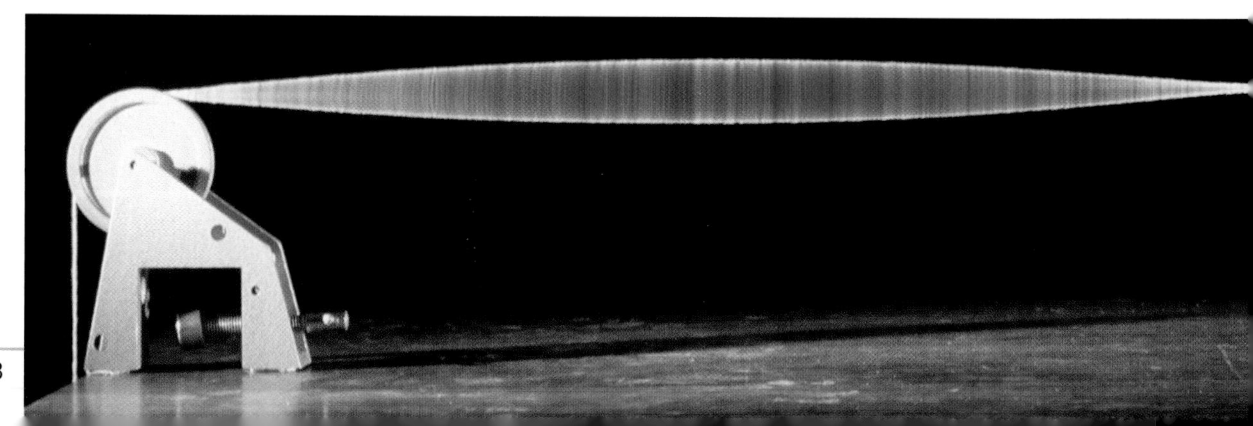

later reinforce each other again. What we get are beats, which can sometimes be heard for the case of interfering sound waves.

STANDING WAVES

What happens when two waves of the same frequency and amplitude are traveling in a medium in opposite directions? If the conditions are right, so that they exactly reinforce each other, the waves will not appear to be moving to the right or the left. We will have what is called a stationary or standing wave. You can, in fact, have standing waves with several loops in them, but there will be a node at the ends in all cases.

You can set up a wave of this type using a rope tied to a doorknob. Again, you will have to move the rope up and down, but you will have to do it continuously, and you will likely have to experiment to find the exact vibrational rate you need for a given length of rope. The reason that you have waves moving in opposite directions in this case is that the waves are reflected when they hit the doorknob; they then begin to move back toward you on the rope, and these waves interact with the ones moving away from you.

Interference pattern in a ripple tank, in which waves from two sources are interacting with each other.

RIPPLE TANK

An excellent way of studying waves is through the use of a ripple tank. It usually consists of a shallow pan of water and one or more vibrational devices for creating waves on the water. You can set up vibrations at opposite ends of the trays and then watch what happens as the two waves move toward each other and collide. An interesting interference pattern can be obtained if the two sources are placed close to each other near the center of the tray. You can clearly see where destructive and constructive interference occurs. The regions of zero amplitude, in particular, stand out.

Standing waves do not only occur in water. They can also happen in a column of air. In this case, when the wave travels down a tube it is reflected back when it reaches the end. All the properties of a standing water wave will be present. There will be a node at a closed end of the tube and an antinode, or loop, at an open end. Standing waves also occur in musical instruments of all types—stringed, brass, woodwind, and even percussion.

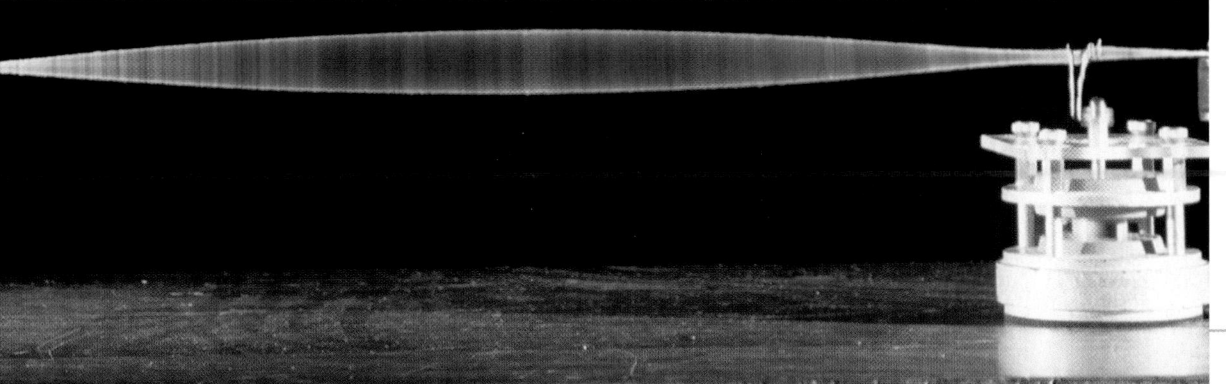

CHAPTER 4

MATTER AND ENERGY

Left: Diagram of an atom. The Greek philosopher Democritus was the first to put forward the theory that all things were made up of incredibly small particles called atoms. Top: Food provides the energy that fuels human activity. Bottom: Energy serves many purposes in our everyday lives. One of the most familiar of these is the powering of electrical devices.

Matter and energy are two of the basic ingredients of the universe. Both have intrigued scientists almost since the beginning of civilization. What exactly are they? From a simple point of view, we can say that matter is the substance of the universe, and energy is what moves it. Indeed, matter is the "stuff" that is all around us; everything we see is composed of matter. We can take a piece of anything, place it on a table, and point to it as "matter." We can't do this in the case of energy. It is not something you can easily describe, mostly because your senses don't detect it directly. You can, of course, see its effects, and everyone knows what the word implies. It is what we need to run our cars, heat our homes, and run the machinery that supplies our everyday needs. Furthermore, it is what makes us feel good and full of vitality, since what we eat gives us energy.

It is safe to say that the study of matter and energy has changed the world. It has taught us many things—some good, some bad. It has brought us comfort and luxuries, but at the same time it has also allowed us to build thermonuclear bombs and other devices of war. For all we have learned, however, many mysteries remain. We may think we have learned everything there is to learn about matter and energy, but this is not so.

Early Understanding of Matter

The first to seriously consider what matter was made of were the early Greeks. Thales of Miletus, who lived in the sixth century BCE, came to the conclusion that water was the basis of all matter. A hundred years later Heraclitus of Ephesus came to an equally strange conclusion: He suggested that fire was the basic substance, and that all matter derived from it.

A radically different approach came from Democritus of Abdera, who lived about 400 BCE. His ideas were so far ahead of his time, however, that most other early philosophers did not take them seriously. He believed that matter was made up of infinitesimally small particles that he called atoms (this is where our word comes from). He suggested that they were indivisible, and that atoms of different materials differed physically. Water atoms, for example, were smooth, whereas atoms of earth were rough and jagged. He even suggested that the motion of these atoms determined some of the properties of the matter.

Aristotle did not agree with Democritus (as we saw earlier, he suggested that everything was made up of four elements: earth, water, air, and fire). Indivisible particles played no part in his theory, and he rejected the idea. Democritus also got little support from other Greek philosophers.

Above: Water being poured into a container of water, creating bubbles below the surface. Water is one of the most common substances on Earth. Top left: The early Greek philosopher Heraclitus concluded that all matter was derived from fire.

Most of the Earth is solid matter.

NEWTON'S IDEAS

For many years after the early Greeks, few speculated on the nature of matter. Democritus's views, however, did not die out, and many people continued to wonder if matter was indeed made of atoms.

Newton was one of them. He defined matter as that which resists change in motion, and also gave considerable thought to its basic structure, finally coming to the conclusion that matter was made up of tiny particles similar to billiard balls that obeyed his laws of motion. Science was not yet ready for a breakthrough in this area; the major problem was that scientists had not identified the "basic substances" of nature. We now refer to them as "elements," but at the time there was considerable confusion about what an element was.

ALCHEMY

One of the reasons little progress was made in understanding the nature of matter after Democritus was that many of the philosophers and thinkers of the time were attracted to alchemy. Alchemy was concerned with the conversion of "impure" base metals into "pure" metals such as gold and silver. Born in Greece and Egypt, it quickly spread throughout Europe, Arabia, China, India, and elsewhere and soon became entrenched in many cultures.

Greek alchemists believed that mercury was a "fundamental" substance and by treating it with sulfur they could change it into gold and silver. One of their major tools was the "philosopher's stone." Alchemists believed that this stone could be used to change base metals to gold. Strangely, it was also believed to cure disease and confer immortality.

Although alchemy was, for the most part, considered to be a black art based on mysticism and the occult, some good did come of it. In the process of experimenting with sulfur, mercury, and various metals using combustion, evaporation, and condensation, alchemists made several important discoveries that were later incorporated into chemistry.

An alchemist at work.

Atoms and Molecules

As we saw, one of the major obstacles in understanding the ultimate nature of matter was establishing which materials or substances were truly elemental. In 1661, however, the English-Irish chemist Robert Boyle (1627–91) published *The Skeptical Chemist,* in which he attempted to describe an "element":

I mean by elements . . . certain primitive or simple or perfectly unmingled bodies; which not being made of any other bodies, or of one another, are the ingredients of which all those called perfectly mixt bodies are immediately compounded, into which they are ultimately resolved.

Boyle was strongly influenced by Galileo and his scientific approach to the problems of nature, and he soon became an ardent practitioner of a scientific method. After many years of experimenting, he was able to identify a number of substances that appeared to be elements. Strangely, though, like many of his contemporaries, he had an interest in alchemy and encouraged others to continue trying to produce gold from other metals. At the same time, with the publication of his book, Boyle broke away from alchemy and helped form the basis of chemistry.

+1
-1
1
H
HYDROGEN
1.00797

John Dalton assigned an atomic weight of one to the element hydrogen.

LAVOISIER

Boyle gave us the definition of an element, but it was Antoine Lavoisier of France (1743–94) who laid down the basic techniques and standards for testing whether a substance fitted Boyle's definition. Early in his career he showed that mass was not gained or lost in a chemical reaction; in other words, the mass of the reactants was the same as that of the final products. But his most important contribution was that he put chemistry on a firm foundation. Chemicals had arbitrary names, and when someone talked about a chemical, no one could be certain what chemical was being referred to. Along with a number of other chemists, he published the book *Methods of Chemical Nomenclature* in 1787. In this book each chemical was given a name based on the elements of which it was composed.

Above: Antoine Lavoisier is said to be the father of modern chemistry. Top left: A portion of the current periodic table of elements.

Two years later, in 1789, he published the first chemistry textbook, titled *Elementary Treatise on Chemistry.* It presented for the first time a unified picture of chemistry and included a list of 23 elements that he had isolated. These two books brought him the title "father of modern chemistry." Sadly, because of his place in society, Lavoisier died by the guillotine during the French Revolution, along with a number of his colleagues.

DALTON

The stage was now set. The definition of an element had been presented, and the idea that matter might be composed of tiny elemental particles had been around for years. The man who brought everything together and formulated the first atomic theory was the English chemist John Dalton (1766–1844).

Dalton assumed that each element was made up of Democritus's "atoms." But he went much further than Democritus. He assumed that all atoms of the same element had the same mass, and different elements were distinguished by being made up of atoms of different mass. In particular, he assigned an "atomic weight" to each of the known elements. Early on he had experimented with various gases, including oxygen and hydrogen, and had determined that oxygen was eight times as heavy as hydrogen. Finding no element lighter than hydrogen, he decided to use it as a reference and assigned it an atomic weight of one.

Dalton went on to define a "compound" as made up of a small number of atoms formed into molecules. Dalton formulated his theory in 1803, and in 1808 he published it in the book *New System of Chemical Philosophy*. His ideas were so logical that they were soon accepted by most scientists. A few, however, held out, and criticized his theory, but Dalton weathered the criticism and in the end was proven right.

John Dalton formulated the idea that matter was made of atoms.

DALTON, THE MAN

Dalton was not your everyday, typical scientist. He started out as a schoolmaster, and as a Quaker he taught at a Quaker school for many years. But he was fascinated by science and soon became a dedicated experimenter.

He set up a crude lab in his home and performed many experiments on gases and other substances. He also had an obsession with the weather; each day for more than 57 years he recorded the temperature, cloud cover, rainfall, and overall weather conditions near his home. He remained a bachelor throughout his life, saying he was so busy that he never had time to marry.

In 1831 he helped establish the British Association for the Advancement of Science, and in 1832 he received an honorary doctorate from Oxford University. Although Dalton spent his life avoiding the spotlight, his funeral was attended by thousands.

Energy

Closely associated with matter is something we introduced earlier, namely work. If you lift a given amount of matter through a certain distance you do work. But to perform this work it takes *energy*. Energy can, in fact, be defined as the ability to do work, or more specifically, it is "realized" work. When we perform work on a mass we set it in motion, which in turn gives it energy. This particular form of energy is called *kinetic energy*.

KINETIC ENERGY

Kinetic energy is energy of motion, so anything in motion (for example, a ball or a car) has kinetic energy. Since motion involves speed or velocity, the formula for kinetic energy will have to depend on velocity. It also seems logical that it will depend on how much mass we set in motion. We therefore have kinetic energy = $\frac{1}{2} mv^2$ where m is mass, and v is velocity. Note that it does not depend simply on v, but rather on v^2. The units of kinetic energy are the same as those of work.

Above: Stopping a moving object causes it to convert energy. Right: When a ball tossed in the air stops its upward movement, its kinetic energy has been converted into potential energy. Top left: Firewood possesses chemical energy.

POTENTIAL ENERGY

Let's perform a simple experiment. Assume we throw a ball straight upward and keep track of its energy. When it is first thrown it has a relatively high velocity and therefore considerable kinetic energy. Gravity is acting on it as it climbs, however, and slows it down. As its velocity decreases, its kinetic energy also decreases. Eventually it stops, and since it has no velocity at this point, its kinetic energy has also disappeared. What happened to it? If you continue watching the ball, you see that it begins to fall, and as it falls it speeds up. This occurs, of course, because gravity is still acting on it.

What has happened is that the original kinetic energy that the ball had is converted to a new type of energy that depends on position, rather than velocity. It is called *potential energy*. At the top, when the ball stops, all its kinetic energy is gone; it has been converted to potential energy. As the ball begins to fall, however, this potential energy is converted back to kinetic energy. In fact, as the ball gains speed, it gains kinetic energy and loses potential energy.

When the ball is just about to hit the ground, all its potential energy is converted to kinetic energy.

The formula for potential energy obviously has to depend on position, or height above the Earth. It is potential energy $= mgh$ where m is mass, g is the acceleration of gravity, and h is the height above the Earth. The units are the same as those of kinetic energy.

An important question at this point is, what happens to the kinetic energy when the ball strikes the Earth? Is it lost? As it turns out, kinetic and potential energy are not the only types of energy, and before we can answer the question we have to become familiar with some of these other forms.

OTHER TYPES OF ENERGY

If you could take a picture of a bat hitting a ball, you would see that the ball was slightly deformed when the bat hit it. Furthermore, if you measured the temperature of the ball when it was deformed, you would see that it had increased slightly. Two new forms of energy are involved in this: deformational energy and heat energy. There are, in fact, several other types of energy, such as chemical energy, nuclear energy, sound energy, and electrical energy. Chemical energy is the energy that is tied up in chemical molecules. It is the energy we get from our food, so it obviously plays an important role in our life. It is also the energy of coal, wood, and oil, so it helps keep us warm. Nuclear energy also now plays a large role in our lives. It is the energy associated with the atomic and hydrogen bombs, but in its controlled form in nuclear reactors it gives us electrical energy to heat and light our homes.

Returning now to our discussion of the ball striking the Earth, we see first of all that it deforms the Earth slightly, and if we could measure the ground's temperature, we would see that it has increased slightly. So energy has not been lost; it has just been changed into other forms.

Prior to this car collision, all of the cars had kinetic energy. During the collision that kinetic energy was converted to other types of energy.

ENERGY IN A CAR COLLISION

One place where energy plays an important role is in the collision of two cars. Of course cars can collide in many different ways, but for simplicity we will assume that the cars collide head-on. Both cars have kinetic energy before the collision, and this energy is converted into other types of energy during the collision. At the moment of collision both cars are deformed, so some kinetic energy is converted to deformational energy. We also hear a loud crash, so some of the kinetic energy goes into sound. And finally the two cars' remnants gain heat, so some of the kinetic energy goes into heat. The amount of deformation, sound, and heating obviously depends on the initial kinetic energy, since all the kinetic energy is transformed into these three forms of energy.

States of Matter

As we have seen, matter can exist in any of three states: gas, liquid, or solid. Indeed, any type of matter can exist in all three forms, depending on its temperature. The reason for this is that regardless of what state matter is in, there are always forces of attraction between the molecules, and these forces determine its form. In solids the molecules are relatively close to one another, and therefore the forces are great enough to hold the molecules in a regular pattern and give matter a definite shape. In liquids the molecules are farther apart, and the forces of attraction are therefore smaller. Because of this, the liquid can maintain a definite volume but can now flow and therefore always assumes the shape of the container it is in. In a gas the distance between molecules is much greater, and the forces are quite weak. Because of this, a gas does not maintain a particular shape or a definite volume. Both its shape and volume assume those of the container.

As you increase the temperature of a given substance, you increase the kinetic energy of its constituent atoms or molecules, and the distance between its molecules increases. It is this increase in separation between molecules that causes the change of state from a solid to a liquid, and finally to a gas.

GASES

Gases have the most kinetic energy and are the most tenuous of the three forms of matter. It might seem, therefore, that they are the least understood of the three forms. But this is not the case: The individual molecules of a gas are so widely spaced compared to those of a liquid or solid that they can be considered to be acting independently of one another. This is particularly helpful in analyzing their properties, and, as a result gases are generally better understood than liquids or solids.

Gases have unique properties. When expanded they cool, and when they are compressed they heat. If you have

Above: Two examples of matter being converted from one form to another: the conversion of a combustible material to smoke (left) and the conversion of water to steam (right).
Top left: A close-up of sulfur crystals.

ever put air in a bicycle tire, you have no doubt noticed this heating. Gases also intermix with other gases, and two different gases diffuse through each other. Open a bottle of perfume in a room and you experience this firsthand; before long you can smell the perfume throughout the room. Gases also flow like liquids, so they can pass through pipelines. This is, of course, how people get the gas they need to run their furnaces and stoves.

One of the most common gases is air, and we see its properties every day. It is a mixture of gases, including nitrogen, oxygen, carbon dioxide, helium, and several rare gases.

LIQUIDS

Liquids are between gases and solids but share more properties with gases than they do with solids. In general they are shapeless and take on the shape of the container they are in. They are so common on Earth (particularly water, which makes up almost four-fifths of Earth's surface area) that it might seem as if they are one of the more common natural substances. In reality, however, aside from water, very few liquids occur naturally on Earth. Two of the only elements that do are mercury and bromine. Yet if you go into a store you see hundreds of different types of liquids on the shelves: oils of various types, alcohol, acetone, ammonia, molasses, and so on. It may surprise you that almost all of these liquids are man-made.

As in a gas, the molecules in a liquid are in constant motion, but they are much closer together than the molecules of a gas. In fact, they are nearly as closely packed as the molecules in a solid, yet they are far enough away from one another so that they can flow. Lowering the temperature of any fluid will eventually transform the fluid to a solid, and raising the temperature will change it to a gas.

A diamond, the hardest material known.

SOLIDS

Most of the naturally occurring elements on Earth are solids. Unlike liquids and gases they have a definite form or shape, and they retain this shape. Because they are easier to handle and retain their form, it might seem that solids would be easier to understand than liquids and gases, but as we saw, this is not the case.

Many of the properties of solids depend on how their building blocks are put together. Change the form of this structure, and frequently you change the properties of the solid. An important breakthrough in the understanding of solids was the discovery that many of them have an orderly arrangement of their atoms, called crystalline structure. Indeed, if the atoms of a solid are arranged in an orderly fashion, it is referred to as crystalline; if there is no order it is called amorphous.

X-rays played an important role in understanding crystals and eventually led to a branch of science called crystallography. When split, crystals break along particular lines or surfaces, and these surfaces usually reflect X-rays in a certain way. Many different geometrical symmetries can be seen in crystals, such as cubic, hexagonal, and rhombic.

Fluids

Technically, both gases and liquids are classified as fluids, but we will restrict ourselves mainly to liquids in this section. One of the first things we think of in relation to liquids is *pressure,* which is the force per unit area exerted by the weight of the fluid. The best way to visualize the pressure that a fluid exerts is to consider the force on a unit area that is at the bottom of a container filled with fluid. If we consider a unit area, the pressure at the bottom is equal to the weight of the column of liquid above it. And the weight of anything, including fluids, depends on its *density,* where density is defined as weight per unit volume. It has units of gm/cm^3 or kg/m^3 in cgs and mks units and lbs/ft^3 in the British system. The density of water, for example, is 62 lbs/ft^3 in British units, 1 gm/cm^3 in cgs units, and 10 kg/m^3 in mks units. Using this in the above definition of pressure we see that the total weight acting on the unit (its pressure) is the density of the water times its height. In a formula this is

$$P = \rho g h$$

where P is pressure, ρ is density, g is the acceleration of gravity, and h is height. The units of pressure are lbs/in^2, $newtons/m^2$, and $dynes/cm^2$.

It is important to point out that the pressure of a liquid at any point is independent of the shape of the container, and the pressure is always perpendicular to the surface we are considering, regardless of how it is oriented.

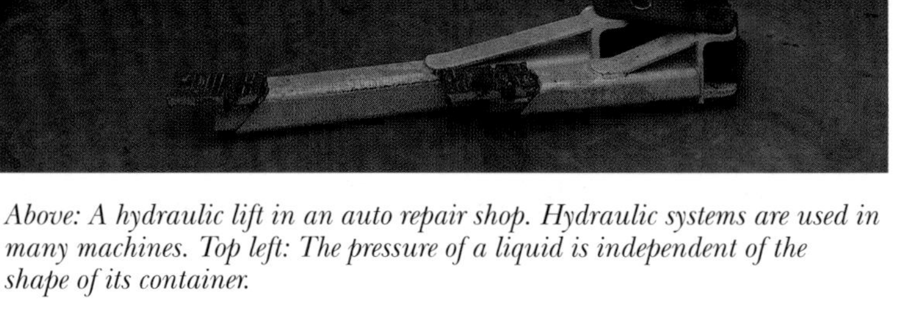

Above: A hydraulic lift in an auto repair shop. Hydraulic systems are used in many machines. Top left: The pressure of a liquid is independent of the shape of its container.

PASCAL'S PRINCIPLE

Another important property of water (or fluids in general)

related to pressure is contained in Blaise Pascal's (1623–62) principle, which can be stated as:

Pressure exerted anywhere on a confined fluid is transmitted unchanged to every portion of the fluid and to all the walls of the containing vessel.

This tells us that it is not only the weight of the fluid that is transmitted to all parts of the fluid as pressure, but that a force applied to the fluid externally is also transmitted in the same way.

Consider a plunger at one point connected to a plunger at a distant point with the same area. If you move the plunger on the left by two inches, the resulting force will be transmitted to the plunger on the right and will push the plunger there outward by two inches.

Furthermore, if the plunger on the right has a larger area than the one on the left, say 10 times as much, the force on it will be increased by 10. It might seem that we are getting something for nothing in this, but we are not. If you look at the movement of the large plunger you see that it moves a much smaller distance than the smaller one. The volume displaced by the two plungers has to be the same, and if the larger one has a greater area, it will move a smaller distance.

This is how a hydraulic jack works, as well as any number of construction vehicles from backhoes to forklifts.

BRAKING SYSTEM IN A CAR

Pascal's principle plays an important role in the braking system in your car. The overall braking system consists of the brake pedal, the master cylinder, the brake lines to the brakes in the wheels, and the brakes themselves at the wheels. When you press the brake pedal, the master cylinder pressurizes the brake fluid, pushing it through the steel tubes that lead to the wheels. The fluid is under pressure, and this pressure is applied to small pistons that lie next to the brake pads. The pistons cause the brake pads to press against a rotor, and the resulting friction stops the car.

The brake drum of a car.

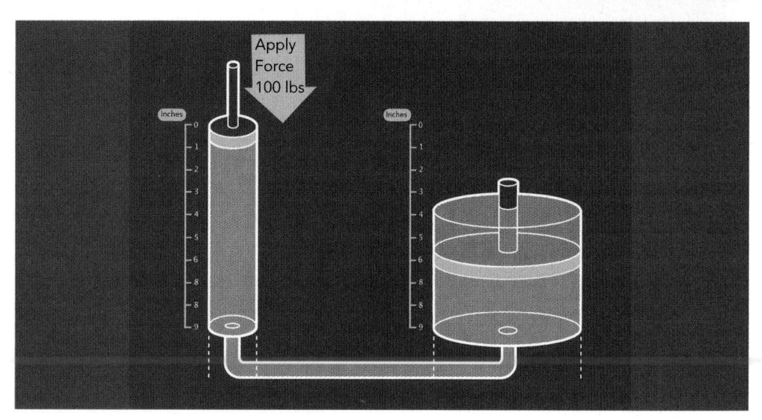

This diagram illustrates a simple hydraulic system with two plungers of different areas connected by a fluid-filled pipe.

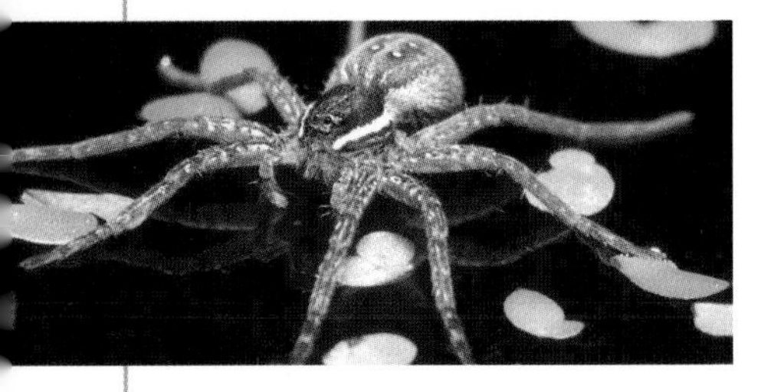

Water Pressure and Tension

Water pressure is exerted not only on the bottom and walls of a container, but against anything within the container. To see the effect of this, consider a solid cube within the container. Each of its six sides will experience a pressure perpendicular to it. The forces on the sides of the cube will cancel one another, but those on the top and bottom won't. Because the bottom is farther down in the water than the top, it will have a greater force on it. The force will, in fact, be greater by the weight of the water the cube displaces, and this means there will be an *upward force on the cube equal to the weight of the water displaced.* This is known as Archimedes' principle.

ARCHIMEDES' PRINCIPLE

Archimedes (287–212 BCE) arrived at his rule when he was asked by the king of Syracuse, Sicily, to determine if a blacksmith had stolen some of the gold he had been given to make a crown by substituting silver for it. (As it turned out he had.)

Archimedes' principle applies to a body that is totally or partially submerged. If the weight of the body is less than the weight of the water displaced, the body will float. This occurs when the density of the body is less than that of water. It is easy to see, therefore, why large ships made of steel float. Most of the interior of the ship is air, and as long as the average density of the steel and air is less than that of water, it will float.

Above: Archimedes discovered his principle of buoyancy while he was having a bath. He is said to have run into the street naked, yelling "Eureka, Eureka, I have solved it!" Top left: The spider is able to stay on the surface of the water because of surface tension.

COHESION AND SURFACE TENSION

A common sight on any pond is small insects gliding across the surface, as if they were skating. You may also have seen an interesting experiment where someone laid a needle carefully on water, and it didn't sink. It is as if there is a "film" on the surface of water stopping the insects and the needle from breaking through—and indeed there is. To see what causes this film we have to look at the forces between the molecules in the water; they are called *cohesive forces.* Cohesive forces exist between molecules in all forms of matter. They are particularly strong in solids, but still fairly strong in liquids. Beneath the surface of a liquid they act in all directions, so there is no net force in any particular direction. At the surface, however, there are no neighboring liquid molecules above, and as a result the forces along the surface are particularly strong. This is what gives us the "film," and it explains why it is more difficult for an object to move through the surface than it is for it to move around once it is completely submerged. We refer to this as *surface tension*. The surface tension of water (at 20°C or 68°F), for example, is 72.8 dynes/cm; for ethyl alcohol it is 22.3 dynes/cm, and for mercury it is 465 dynes/cm.

We can also think of this surface tension as surface energy. An interesting feature of it is that, left to itself, it will reduce a surface to its minimum area. This is why a small quantity of liquid suspended in air will take the shape of a sphere (raindrops, for example); a sphere has the smallest surface area for its volume.

As we saw, if all the molecules are the same, the intermolecular forces are cohesive forces. But there are also attractions between unlike molecules—for example between water molecules and the molecules of a glass tube containing the water. These are referred to as *adhesive forces*. One of the results of this is *capillary action*. If you put a narrow glass tube in a container of water, you will see water

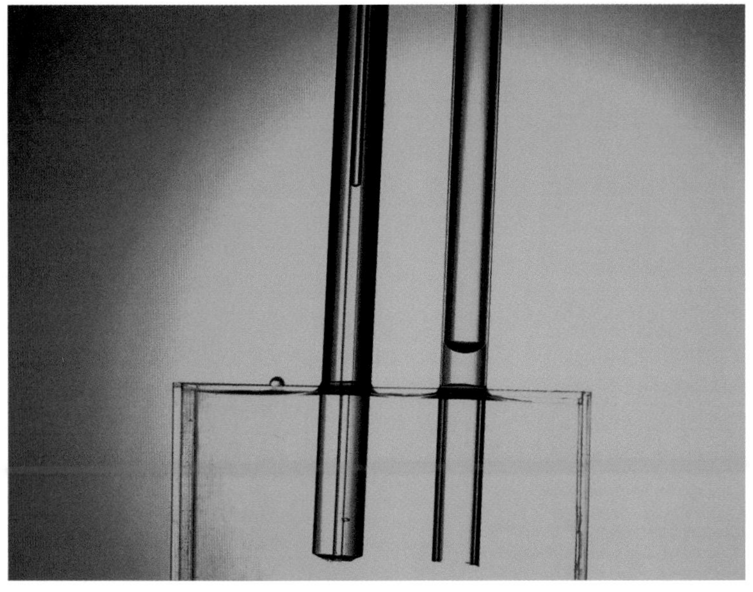

Left: The capillary action of water draws it up a glass tube. Water rises to different levels in glass tubes of different widths. Above: Archimedes' principle explains how steel ships are able to float on water.

rise in the tube. In this case the attractive forces between the water and the glass are greater than the cohesive forces within the water, and the water rises upward to increase the water-glass contact. As it rises, however, gravity acts on it and pulls it back down. Eventually the two forces balance, with the water in the tube at some point above the surrounding water.

Heat and Thermal Energy

Above: Portrait of two British physicists, Lord Rayleigh (at left) and Lord Kelvin. Lord Kelvin, born William Thomson, proposed the existence of an absolute zero temperature. Top left: When you dribble a basketball, the ball heats the ground where it hits.

As we saw earlier, heat is a form of energy. A ball striking the Earth heats the ground where it hits. There is, in fact, an exact relationship between heat and energy (or equivalently, work), which was discovered by Count Rumford, an American-born British physicist. As a major general in the Bavarian army, he was put in charge of manufacturing brass cannons, and he personally supervised the boring of the cannon barrels. He was amazed by the amount of heat produced and decided to measure it. To do this he cast a specially shaped cannon barrel that could be insulated against heat loss. He then immersed the drill and barrel in a tank of water and measured the increase in temperature of the water as the boring was going on. From it he determined how much heat was produced. But of particular importance, he also made a calculation of how much heat was produced for a given amount of mechanical work. We now refer to this as the *mechanical equivalent of heat.* The value he obtained was too high, but it was a first step, and about 50 years later James Prescott Joule of England obtained a more accurate value of 4.18

joules/cal (see page 67 for the definition of a calorie). What this tells us is that it takes 4.18 joules of energy or work to produce 1 calorie of heat.

TEMPERATURE SCALES

To measure the amount of heat absorbed or emitted by an object, we have to know its temperature before and after the heat exchange took place, and the device we use to measure temperature is the thermometer. The most common thermometer uses mercury in a glass tube. Mercury is ideal because its freezing point is considerably lower than that of water, and its boiling point is much higher than that of water. Furthermore, it expands uniformly with temperature change.

Three things are necessary in setting up a temperature scale: You must have two reference points, and you have to fix the size of the unit of the scale. It was found early on that the freezing and boiling points of water were ideal as reference points. The first scale based on them was set up in 1714 by the German physicist Gabriel Fahrenheit. He took the reference points to be 32 degrees for freezing and 212 for boiling; between these two points were 180 degrees (which of course determined the size of the unit). This is now referred to as the Fahrenheit scale.

In 1742 Swedish scientist Anders Celsius set up another scale. He selected the freezing point of water to be 0 degrees, and the boiling point to be 100 degrees. In many ways this scale made more sense, but strangely the Fahrenheit scale eventually came into more extensive use in the Western world. Celsius's scale was originally called the centigrade scale, but in the 1950s it was renamed the Celsius scale.

Since we have two scales, both of which are used extensively, it is important to be able to convert between them. The simplest way is to note that there are 9/5 (180/100) Fahrenheit units in one Celsius unit, and that 0 degrees on the Celsius scale is 32 degrees on the Fahrenheit scale. Using this, we have the formula

$$F = (9/5)C + 32$$

Another way to obtain this is to draw the two scales, labeling the freezing and boiling points of each scale. Then take the ratio

$$\frac{C - 0}{100 - 0} = \frac{F - 32}{212 - 32}$$

This gives the same result as above and is easier to remember.

ABSOLUTE ZERO

Earlier we described the Celsius and Fahrenheit scales, but there is actually a third scale of importance. It was noticed early on that as temperature goes down, the volume of a gas contracts, and that you could calculate the temperature at which the volume becomes zero. This occurs at -273°C. What was the significance of this temperature? In 1848 Lord Kelvin of England pointed out that it appeared to represent the lowest temperature possible, in other words, *absolute zero*. If this was the lowest possible temperature, it made sense to define a new temperature scale using it. The lowest temperature on this scale (that is, zero degrees) would be -273 degrees on the Celsius scale. We now refer to this scale as the absolute or Kelvin scale.

Count Rumford, Benjamin Thompson, was one of the first to question caloric theory. His work led him to discover the relationship between heat and work.

Linear Expansion, Specific and Latent Heat

When any substance is heated, the atoms vibrate more vigorously, and therefore they move away from one another. This causes the overall volume of the substance to increase. If we consider only one dimension of this expansion it is referred to as *linear expansion*. Expansions of this type are particularly important in construction, as metal rails, for example, can expand considerably if the temperature changes by a large amount, and this expansion must be allowed for. In order to do this we need what is called the coefficient of linear expansion (usually designated as $\boldsymbol{\alpha}$). It is the relative increase in length for a unit temperature increase. To illustrate it, assume we have a metal rod of one meter. Using the symbol $\boldsymbol{\Delta}t$ to represent the change in temperature, if we increase the temperature of this rod by $\boldsymbol{\Delta}t$, its new length will be $1 + \boldsymbol{\alpha}(\boldsymbol{\Delta}t)$. This applies, of course, to any length L, so we can write

Above: An expansion joint of a bridge prevents damage that might result from thermal expansion of the bridge structure. As the bridge is heated by the Sun, the components tend to expand. If the bridge was totally rigid, the components could buckle or deform. The joint allows some expansion, and the interlocking teeth make it safe to drive over. Top left: As water molecules heat up, they move apart.

$$L = L(1 + \alpha\Delta t)$$

We talked about heat earlier without formally defining it, except for saying it was a measure of the internal energy of the molecules making up the material. To define it mathematically, we have to begin by defining the unit of heat, namely, the *calorie.* A calorie is the quantity of heat required to raise the temperature of one gram of water from 14.5°C to 15.5°C. In the British system we have the Btu, which is the amount of heat required to raise the temperature of one pound of water by one degree Fahrenheit. One Btu is 252 calories.

SPECIFIC HEAT

To define heat more accurately, consider a simple experiment. Assume we have a small block of copper and a container of water. Assume further that the water is at 20°C and that we heat the block to 100°C and then drop it into the water. What we observe after a few moments is that the water has increased slightly in temperature and the copper block has decreased in temperature. In fact, they are now both at the same temperature. This means the block has lost heat and the water has gained heat. We could, in fact, calculate what this final temperature is if we knew how much heat is needed to raise the temperature of one gram of the copper by one degree. This is called its specific heat.

The specific heats of all substances are now well known. For copper it is .093 cal/gm°C. Specific heats are usually designated by c. With this we can now give a formula for heat

$$H = m(\Delta t)\, c$$

where H is heat, c is specific heat, and Δt is the temperature change. All kinds of heat problems can now be solved using this definition. The only thing we have to remember is that the total heat of all the components of the system before the heat reaction has to be equal to the total heat after the reaction.

LATENT HEAT

The above procedure, called the "method of mixtures," works in all problems, as long as you do not exceed the freezing and boiling points of water (or other liquid). Consider what happens when you do. Assume we have a bucket of water and ice and we heat it. Surprisingly, we find that the water stays at 0°C until all the ice is melted; indeed, not until all the ice is gone does the temperature of the water start to increase. It is easy to show, in fact, that 80 calories of heat must be supplied to melt each gram of ice, and no change in temperature takes place while this heat is absorbed. This delay in temperature change is called the *latent heat of fusion.*

The same principle applies with the evaporation of water to steam. Not until all the water is vaporized will the temperature of the steam increase. In this case we have the *latent heat of vaporization*; it is 539 cal/gm. In practice we can still use the method of mixtures described above, but we must take latent heat into consideration if we cross the freezing or boiling point of water (or other liquid).

COUNTING CALORIES

Most people are familiar with calories in relation to the food they eat; these calories are a measure of how much energy is produced from a given amount of food. Strangely, the nutritional calorie is not the same as the calorie used in physics. When nutritionists refer to a calorie, they mean a kilocalorie (1,000 calories).

You use calories when you run or work, but the calories from food are not the same as the calories associated with physics.

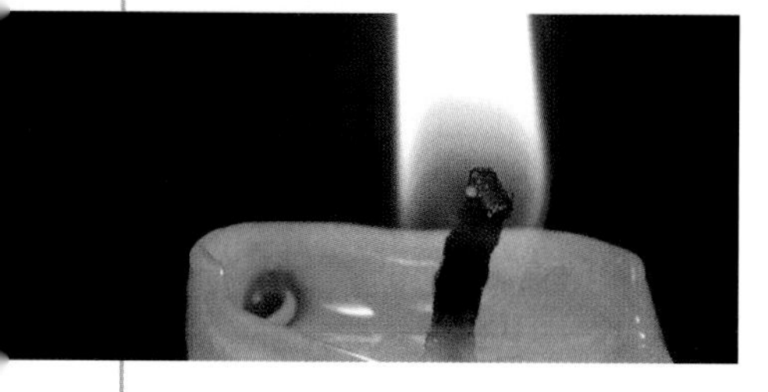

Heat Transfer

Heat is like water in many ways. Water flows from a high point to a low point; heat flows from a point where the temperature is high to where it is low. Furthermore, the greater the difference in height, the faster water flows, while the greater the temperature difference, the greater the rate at which heat flows. Heat can, in fact, be transferred in three different ways: *conduction, convection,* and *radiation.* Heat flows via conduction when there is transfer of energy as a result of collisions in neighboring molecules. Heat flows via convection when there is mass transfer of a heated substance. And heat is transferred via radiation through what is called radiant energy.

Above: The transfer of heat by convection is present in Earth's atmosphere. Top left: A candle flame emits radiant energy.

Wood does not conduct heat and is therefore a good material for cooking utensils used to stir hot food.

CONDUCTION

A good example of heat transfer by conduction occurs when you heat one end of a long metal rod in a flame. As the end in the flame heats up, the atoms gain kinetic energy and vibrate more energetically, and as they do so their motions extend farther and farther from their equilibrium positions. This causes them to collide with neighboring atoms, and as these neighbors begin to vibrate they in turn disturb their neighbors. Kinetic energy is therefore transferred from one end of the bar to the other. How fast this energy, or heat, travels depends on three variables. The first is called the *thermal conductivity* of the metal. It is usually designated as k, and has units of cal/(cm^2 sec)(°C/cm) in cgs units and Btu/(ft^2 hr)(°F/in) in British units. Thermal conductivities can vary considerably. Silver, for example, has a very high conductivity of .99 cal/(cm^2

sec)(°C/cm); copper is also high at .92 cal/(cm^2 sec)(°C/cm). Glass, on the other hand, has a low conductivity of .0025, and air has a conductivity of only .000053. When the conductivity is very low, we refer to the material as a thermal insulator. The second variable is called the *temperature gradient*, which is the change in temperature over a given length of the rod: The greater the gradient, the faster the rate. Finally, the rate the heat travels also depends on the cross-sectional area of the bar.

The insulation of a house is important to reduce heat loss.

INSULATION

Convective currents have to be taken into consideration when selecting the type of insulation to be used in buildings. It might seem that it would be best, for example, to leave the space between the outer support beams empty (filled only with air), since air has one of the lowest known thermal conductivities. But if we do this, convection currents can be set up in the empty space that can lead to considerable loss of heat. Because of this it is best to break the space down to small, isolated spaces where convection currents cannot occur. The best insulation is therefore spun fiberglass, rock wool, or other materials of this kind. They are both poor conductors (good insulators), and they do not allow convection currents to occur.

CONVECTION

The transfer of heat by convection occurs in both liquids and gases. It is generally associated with pressure differences brought about by changes in density. If heat is applied, for example, to the bottom of a container of fluid, the heated fluid expands and becomes less dense as the vibrating molecules move apart, and as a result it moves upward.

Convection occurs in our atmosphere and in the ocean, and it is used in heating houses. It can be classified as free convection or forced convection. Free convection occurs when the motion of the fluid is due entirely to the temperature differences within the fluid. Forced convection, on the other hand, occurs when the motion is a result of an external influence, such as a pump or a fan.

RADIATION

Heat transfer by radiation is different from the above two processes in that it does not require a medium. Energy emitted by the heated filament of a lightbulb, for example, reaches us even though there is a vacuum in the space around the filament. Energy from the Sun also reaches us through the vacuum of space.

Even though there is no medium between us and the source, the "radiant energy" we receive vibrates the atoms in our skin or body in the same way it would if the transfer medium were conduction or convection. Radiant energy is, in fact, emitted by all bodies that have a temperature greater than their surroundings.

CHAPTER 5

THERMODYNAMICS

Left: The thermodynamics of black holes is now an important area of research in physics. The photo shows the region at the center of our galaxy (the Milky Way) where a huge black hole known as Sagittarius A resides. It has a mass three million times that of our Sun. Top: Thermodynamics applies to all heat engines, including jet engines. Bottom: Thermodynamics is applicable to many natural phenomena, including our atmosphere and the storms that occur within it.

Thermodynamics deals with the relationship between heat and work. More explicitly, it is concerned with the conversion of mechanical energy into thermal energy, and the reverse process, the conversion of heat to work. Because the laws of thermodynamics are independent of the atomic makeup or structure of a system, they provide a particularly powerful tool for scientists. They allow predictions to be made about matter and help us understand many of the interconnections among the physical observables of nature.

Thermodynamics was first applied to steam engines, and it does indeed tell us a lot about them, but its principles are so all-encompassing that it can be applied to many other processes and problems in both science and engineering. A few examples are: engines of all kinds, chemical reactions, the atmosphere, black holes, transport phenomena, and the universe itself. The basics of thermodynamics are now an integral part of many different sciences, including chemistry, biology, chemical engineering, cellular biology, material science, meteorology, and others. It is difficult to overstate its usefulness.

Closed Systems and Conservation

Thermodynamics is the branch of physics that deals with the movement or flow of heat and the relationship between heat and work. Of fundamental importance in this study is what is called a *system,* which is defined as a certain amount of matter (for example, a gas or a fluid) enclosed by a surface. This surface need not be fixed in shape or volume and may be imaginary. We could, for example, be considering a gas in a cylinder with a piston at one end that can compress it. The region beyond the surface is usually called the environment.

Three types of systems are distinguished:

Closed systems, which are able to exchange energy (work or heat) but not matter with the environment outside the surface. A good example of such a system is a greenhouse; heat is exchanged with the environment, but not work.

Open systems, which can exchange energy (heat or work) and matter with their environment. A good example of an open system is the ocean.

Isolated systems, which are completely isolated from their environment. They do not exchange heat, work, or matter with it.

We are interested in the *state* of the system, which is defined by certain variables such as pressure, density, temperature, and how it changes. Of particular interest are systems in which there is a state of equilibrium.

Above: The law of conservation of energy tells us that energy cannot be created or destroyed, but it can be converted from one form to another. In the above device radiant energy is being converted to electrical energy. Top left: A closed system is one that does not exchange matter with its environment, such as a greenhouse.

CONSERVATION OF ENERGY

In the above systems all considerations of heat flow must assume that the heat will not suddenly vanish to nothing, or rise out of nothing. This is entailed in the *principle of the conservation of energy.* We have to some degree already encountered it. As we saw earlier, energy can take on many forms; we looked at kinetic and potential energy in considerable detail. In fact, we saw that when a ball is thrown upward, kinetic energy is transformed to potential energy, and when it falls, this potential energy becomes kinetic energy again. This seems to imply that energy cannot be destroyed or created out of nothing; it can only be changed in form. This is the principle of the conservation of energy.

Some of the other forms of energy, as we saw earlier, are chemical energy, electrical energy, and sound energy.

REVERSIBLE AND IRREVERSIBLE PROCESSES

Two types of processes are important in thermodynamics: reversible and irreversible. Any

Left: The ocean is a good example of an open system. Below: Thermodynamic processes can be reversible or irreversible. An irreversible process is one in which the system and its surroundings cannot return to their original conditions, as in an explosion.

process that can be made to go in the reverse direction by a series of infinitesimal changes is referred to as reversible. A good example is water in an insulated container. If the water is heated to form steam, for example, and then the temperature is lowered so that condensation occurs, energy is returned to the heater, until finally everything is the same as it was originally.

Any process that is not reversible is irreversible. Irreversible processes usually occur suddenly; an explosion, for example, is irreversible. If the process involves friction or electrical resistance, it is also irreversible.

Cycles are also of particular importance in thermodynamics. A cycle is a succession of changes in the system—usually changes in pressure, volume, and temperature—that end with the system in its original state. Over the next couple of sections we will discuss several cycles of importance.

Ideal Gases

Gases play an important role in thermodynamics, but real gases are difficult to deal with, so we approximate them with what we call an *ideal gas.* An ideal gas is one in which the collisions between the molecules are perfectly elastic, and there are no forces between individual molecules; the molecules can, in effect, be visualized as billiard balls in collision. All their energy is kinetic, and we can associate a temperature with this energy.

An ideal gas can be characterized by three variables: pressure (P), volume (V), and temperature (T). The relationship among these variables is given in what is called the ideal gas equation. Before we get to it, however, we should look at two early gas equations. The first was formulated by Robert Boyle in 1662. He considered the relationship between pressure and volume when temperature is held constant and found that

$$PV = \mathbf{k}$$

where **k** is a constant. This tells us that the product of pressure and volume remains constant whenever a system undergoes a change. It is known as Boyle's law.

Jacques Charles (1746–1823) gave us another equation in late 1800. He did not publish his results, however, and they were rediscovered by Joseph Louis Gay-Lussac in 1802. Charles considered the relationships between temperature and pressure and volume and temperature where

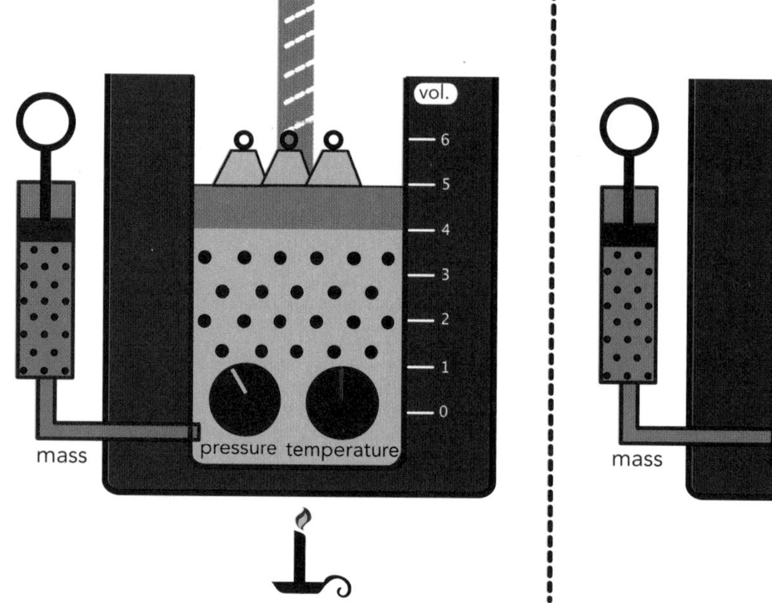

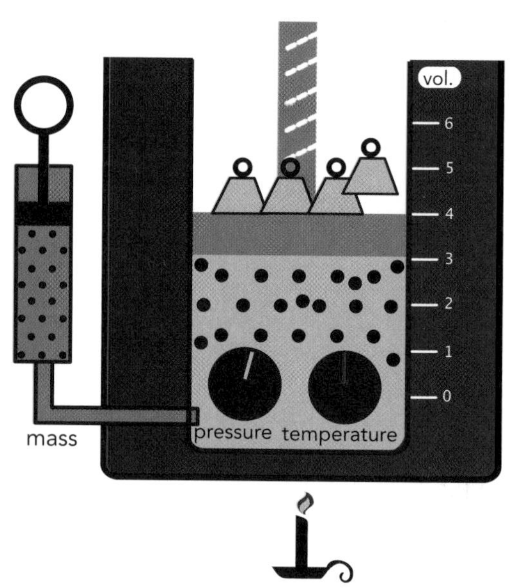

Above: Boyle's law states that for a given mass at constant temperature, the pressure times the volume is a constant. This diagram shows that when we exert pressure on a given amount of gas (shown in yellow), the volume of the gas will decrease, as seen on the right. Top left: Billiard balls in collision help to visualize the collisions between molecules in an ideal gas.

1

T is the absolute temperature, and found

$$P/T = \mathbf{k}_1 \quad \text{and} \quad V/T = \mathbf{k}_2$$

where $\mathbf{k}_1$ and $\mathbf{k}_2$ are constants. These two expressions are known as Charles's law (or sometimes as Gay-Lussac's law).

The two above laws can be combined into what is called the ideal gas equation

$$PV = n\mathrm{R}T$$

where R is the gas constant, which has a value of 8,317 joules/(kmole)(°K), and n is the number of kilomoles (1,000 moles) of the gas, where a mole is the mass of the gas (in mks units) divided by its atomic mass.

PV DIAGRAMS

A useful diagram for dealing with gases is a plot of pressure versus volume, which is known as a PV diagram. The PV diagram of an expanding gas is shown in the figure at right. In this case we are assuming that the gas is in a cylinder with a plunger (or piston) at one end that is moved outward, causing the gas to expand. In the diagram we represent this by the line from point 1 to point 2.

We have to be careful in such a system, however, as we have not specified how the expansion takes place, and there are several possibilities. If the gas expands without changing its temperature, it is called an *isothermal expansion.* In this case the system must be in contact with a heat reservoir, so that heat flows from the reservoir as the gas expands. (Without this reservoir, the temperature of the gas would drop.)

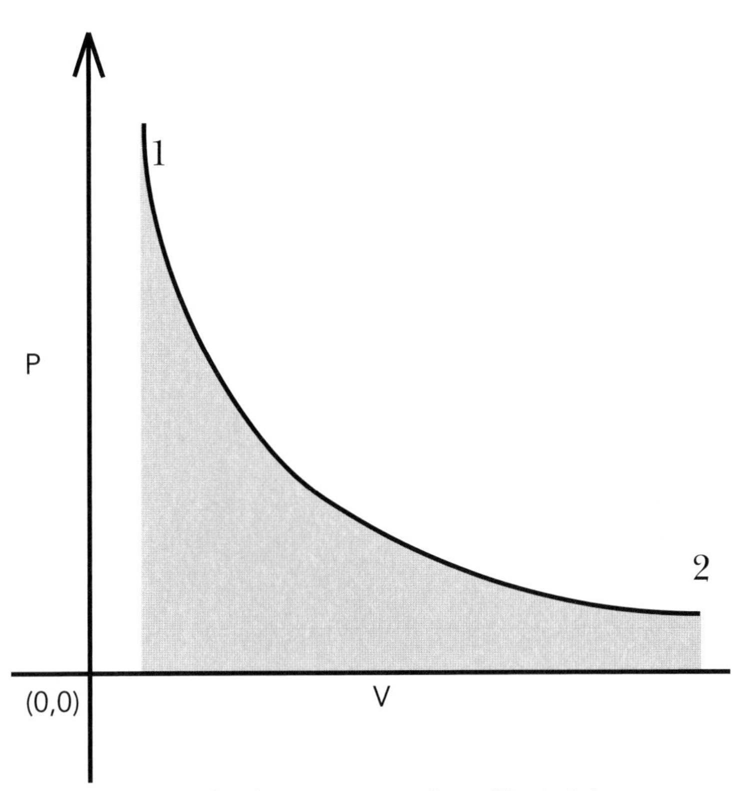

A PV diagram is a plot of pressure versus volume. The shaded area represents the work done by an expanding gas when they system is changed from 1 to 2.

A second case of interest is an *adiabatic expansion.* This occurs when there is no change in the heat of the system. For this, the system must be thermally insulated from its environment, so no heat is taken in from it, or lost. Work is done in both adiabatic and isothermal systems.

OTHER CASES OF INTEREST

Several other cases are also of interest. If the pressure is constant during the expansion, we refer to it as an *isobaric process.* A good example is water in the boiler of a steam engine; if it is heated to its boiling point, vaporized, and the steam is superheated, each of these processes is isobaric. In the same way, if the volume is held constant and a change is made, it is referred to as an *isometric process*, or sometimes as an isovolumetric process. Other variables, which we will introduce later (such as entropy), can also be held constant. All of these processes can be represented in diagrams.

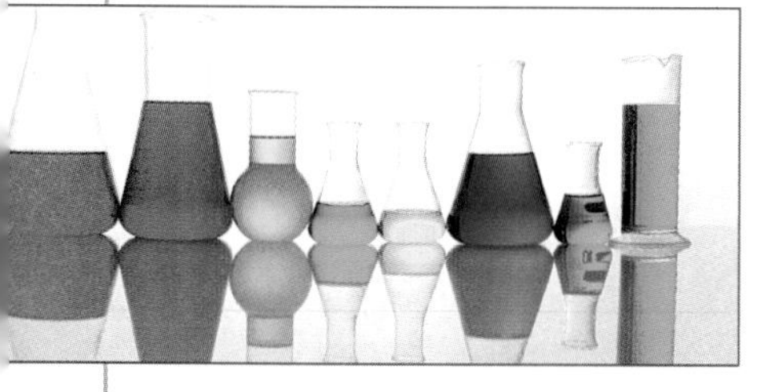

First Law of Thermodynamics

To explain the first law of thermodynamics we have to begin by considering a *thermo-dynamic system*. As we saw earlier, such a system can be open or closed. In an open system we consider both the system and its surrounding environment. This environment can contain one or more heat reservoirs (which are large enough so that their temperature is not affected when heat is withdrawn) and heat sinks. A good example of a thermodynamic system is a gas enclosed in a cylinder with a piston at one end that allows it to change the volume of the gas.

A system is specified by its pressure, volume, and temperature, and also by its physical makeup. We are particularly interested in systems that are in equilibrium; this occurs when the gas has the same values for P, V, and T at all points throughout the gas. Specification of these variables gives the *state* of the system, and an equation called the *equation of state* links the three variables and tells us what happens when the system changes. In short, when external conditions are changed, the thermodynamic system changes its state, and P, V and T adjust to a new equilibrium.

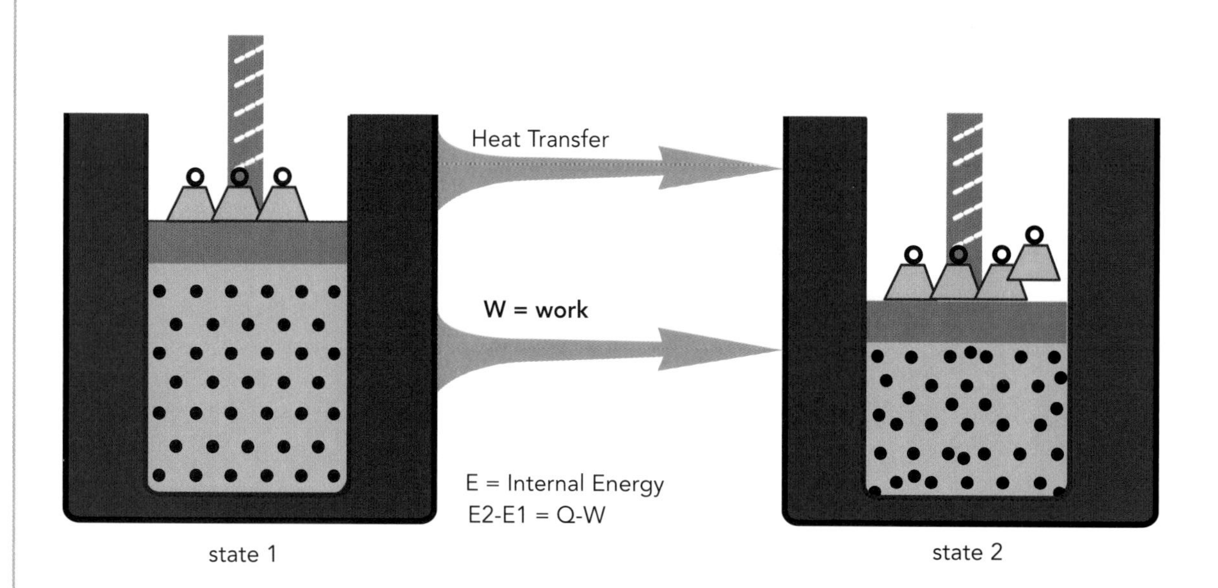

Above: The initial state (state 1) of a system in equilibrium is shown on the left. If it is compressed to equilibrium state 2 (right), a certain amount of heat, or Q, will be transferred, and an amount of work, W, will be done by the system. The difference in internal energies of the two states is given by Q-W. This is known as the first law of thermodynamics. Top left: A system can be anything—a piston, a living organism, or a solution in a test tube.

The first law of thermodynamics applies to the weather. A change in temperature plus a change in air pressure causes a change in the weather.

WORK, HEAT, AND INTERNAL ENERGY

The first law of thermodynamics deals with work, heat, and internal energy, so it is best to review each of these concepts. Assuming we are dealing with a gas, heat is the energy associated with the molecules of the gas. More generally we can say that heat is the energy that is transferred from body to body. The total energy content of a body is referred to as its *internal energy*, and a good measure of this energy is its temperature. Mechanical work done on a body also increases its internal energy; this occurs, for example, when a gas is compressed. Furthermore, when a gas expands, internal energy does mechanical work.

With this we are now in a position to state the first law of thermodynamics and give its formula. In its simplest form it is: *When heat is transferred into other forms of energy, or conversely when other forms of energy are converted to heat, the total amount of energy is constant.*

In terms of work, heat, and internal energy we can state it as: *If the state of a system is changed by applying heat or work (or both), the change in energy of the system must equal the energy applied.* In a formula this is

$$\Delta E = Q + W$$

where ΔE is the change in energy, Q is the added heat, and W is the work done on the system. This law applies only to closed systems, and it is easy to see that in its generalized form it is nothing more than the conservation of energy. The conservation of energy, in fact, forms the foundation of thermodynamics.

British physicist James Prescott Joule, in whose honor the unit of energy is named.

JAMES PRESCOTT JOULE

Several of the breakthroughs that led to the first law of thermodynamics were made by James Prescott Joule. Joule was the son of a wealthy brewer in Lancashire, England. His early education consisted mostly of home tutoring, but he soon became interested in science and set up a home laboratory. As a young man, he studied under Dalton for a short time and was strongly influenced by Dalton's ideas on atoms. Joule became interested in the nature of heat and its relationship to mechanical work in the 1840s. Through experimentation, Joule obtained a value for the mechanical equivalent of heat (J) of approximately 4.15 joules/cal (which is close to our presently accepted value). Joule announced his discovery to the scientific community in 1847, but few were interested. Over the next few years Joule gave a series of lectures on the topic and caught the attention of William Thomson (Lord Kelvin). The two men soon began working together, and within a couple of years they showed that when a gas expanded freely, its temperature dropped. This is now referred to as the Joule-Thomson effect.

Carnot and the Heat Engine

In the 1820s the French physicist Sadi Carnot (1796–1832) became interested in how much work a heat engine, which is an engine that converts heat energy to mechanical work, could do. He was mainly interested in steam engines, but the principles he developed apply to all heat engines. At the time he began working on the problem, steam engines were notoriously inefficient; about 95 percent of the heat energy of the burning fuel was being wasted. Carnot was sure this efficiency could be improved.

In a steam engine, heat flows from a hot region (the steam cylinder) to a cold region (the condenser); in other words, the engine is supplied with energy in the form of heat at a high temperature (T_1), the engine performs work, and then it rejects heat at a lower temperature (T_2). Carnot disregarded the physical details of the steam engine and concentrated on these two temperatures. He visualized what is now called the Carnot engine, an idealized engine in which no losses occurred due to friction or radiation. Its efficiency, which is greater than

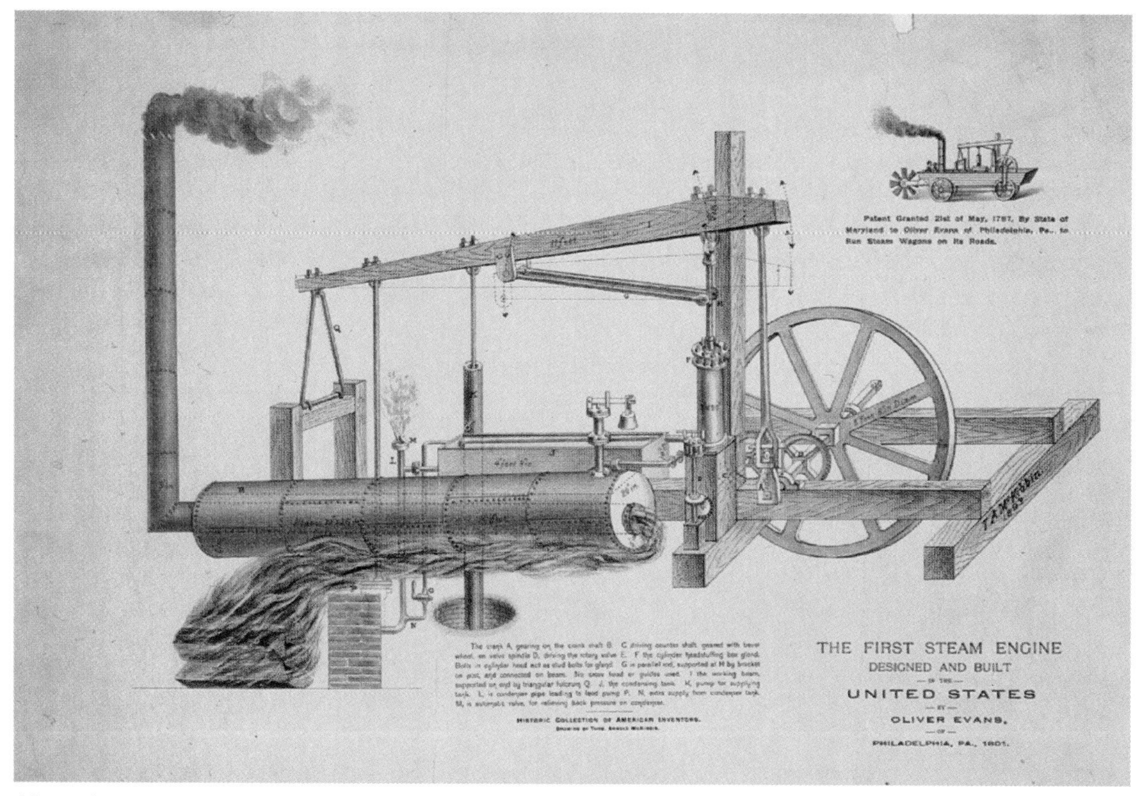

Above: A steam engine. This one was built in the United States by Oliver Evans of Philadelphia in 1801. Top left: Diesel engines, like those found in tractor-trailer trucks, are more efficient than gasoline engines.

that of any other heat engine, is given by

$$\text{Efficiency} = 1 - T_2/T_1$$

It is evident from this equation that if we want to maximize the efficiency, we have to make T_2, the exhaust temperature, as low as possible, and T_1, the engine temperature, as high as possible. In short, the greater the ratio of the temperatures T_1 and T_2, the higher the efficiency. Furthermore, since the temperatures are in degrees absolute (Kelvin), for 100 percent efficiency we would need T_2 to equal 0°K, and as we will see later, this is not possible.

THE CARNOT CYCLE

We can visualize a hypothetical Carnot engine as a simple cylinder with a freely moving piston. The cylinder is filled with an ideal gas.

We can summarize the operation of the Carnot engine as follows: A definite quantity of heat is absorbed from a reservoir at a high temperature; part of this heat is converted to useful work, and the balance is expelled into a low-temperature reservoir. The latter heat is considered to be wasted heat. Many engineers tried to create this perfect cycle for 100 percent efficiency of their engines. Rudolf Diesel (1858–1913) had the most success with this. After 13 years of hard work, he developed a near-perfect engine that ran on its own power for the first time in 1893.

Sadi Carnot, the French physicist and engineer who is generally credited with being the founder of the science of thermodynamics.

WILLIAM THOMSON (LORD KELVIN)

Carnot's work was not recognized immediately; in fact, it took 20 years before anyone began to pay attention to it. Lord Kelvin of Scotland was one of the first to take it seriously. For several years he had been concerned with the strange relationship between heat and work: Work could be converted to heat without any complication (indeed, all of a given amount of work could be converted to heat), but this was not the case when heat was converted to work. Some heat was always "wasted" as the system passed from a high temperature to a low temperature. Kelvin thought of this as a "dissipation of energy." He began studying Carnot's work in the late 1840s and was soon convinced that two different temperatures were needed in a heat engine; in effect, it could not operate at a single temperature. This was an important step. He also helped others understand the problems associated with heat engines more fully by introducing the absolute temperature scale. It was soon shown to play a critical role in thermodynamics.

Gasoline engines undergo what is called the Otto cycle.

THE OTTO CYCLE

Another cycle that is important in thermodynamics is the one associated with the combustion or gasoline engine. It is difficult to analyze the standard gasoline engine accurately, so we are forced to make a number of approximations. The idealized cycle we use is called the Otto cycle in honor of Nikolaus Otto (1821–91), the inventor of the four-cycle engine. In this cycle an air-gasoline mixture is taken in, compressed by a piston, a spark plug is fired, and the resulting explosion creates pressure that forces the piston back, giving the engine its power. An exhaust valve is then opened and the exhaust expelled.

Second Law of Thermodynamics

The second law of thermodynamics applies to a closed system and can be stated as:

It is impossible to construct an engine that, operating in a cycle, will produce no effect other than the extraction of heat from a source and the conversion of this heat completely into work.

There are several different ways to state the second law, but they are all equivalent. (Another way that is easy to visualize is that heat will not spontaneously flow from cold temperatures to hot temperatures.) One of the things this law showed was that several types of "perpetual motion" machines were impossible. Perpetual motion machines are machines that run indefinitely without an appropriate source of energy. An example is a ship that propels itself by using heat from the water—in other words, by taking in water at a certain temperature and expelling it at a lower temperature. By the second law this is impossible.

CLAUSIUS AND ENTROPY

The above statement of the second law applies mainly to heat engines. But as we saw earlier, there are many other forms of energy, and because of this there was a need for a generalization of the law. This was accomplished by the German physicist Rudolf Clausius (1822–88). He introduced a new quantity he called *entropy*, which allowed him to formulate the law in a different way. Entropy is a rather abstract concept that cannot be measured directly, but we can get a good idea of what it is by considering a reversible process that is occurring very slowly. A change of entropy is then the sum of the small amounts of heat absorbed at each part of the process divided by the absolute temperature at that point. In terms of a formula it is

$$\Delta S = \Delta Q/T$$

where ΔS is the change in entropy, ΔQ is the change

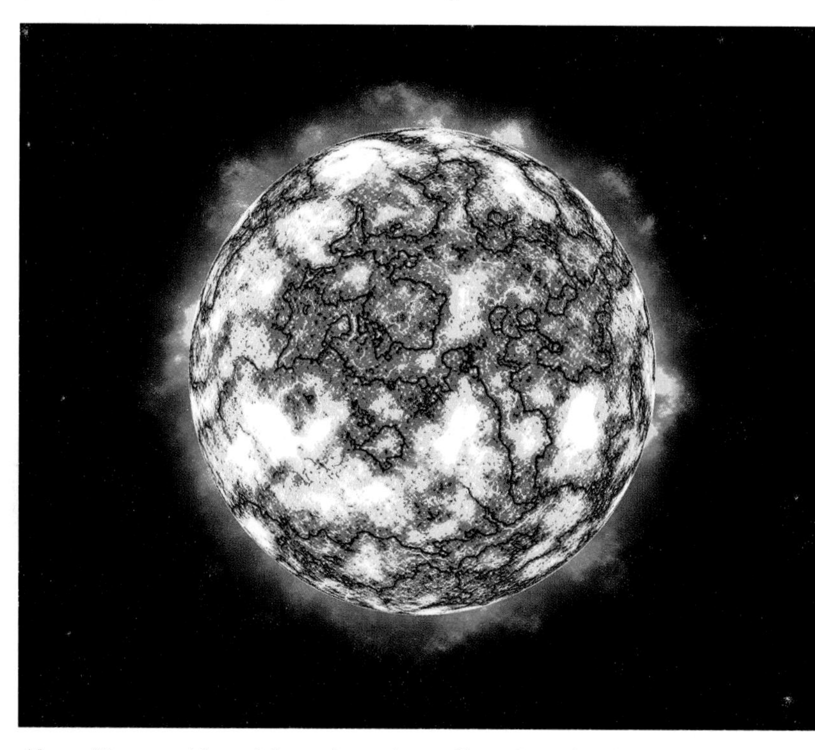

Above: The second law of thermodynamics applies only to closed systems. But no system can be entirely unaffected by outside influences, such as the Sun. Top left: Melting ice is a classic example of entropy increasing.

Above: Rudolf Emmanuel Clausius, the German theoretical physicist who introduced the concept called entropy, which is the ratio of heat content to absolute temperature.

Below: A perpetual motion machine is a machine that continues to work indefinitely without a source of energy. No such machine actually exists. Perpetual motion machines violate either the first or second law of thermodynamics.

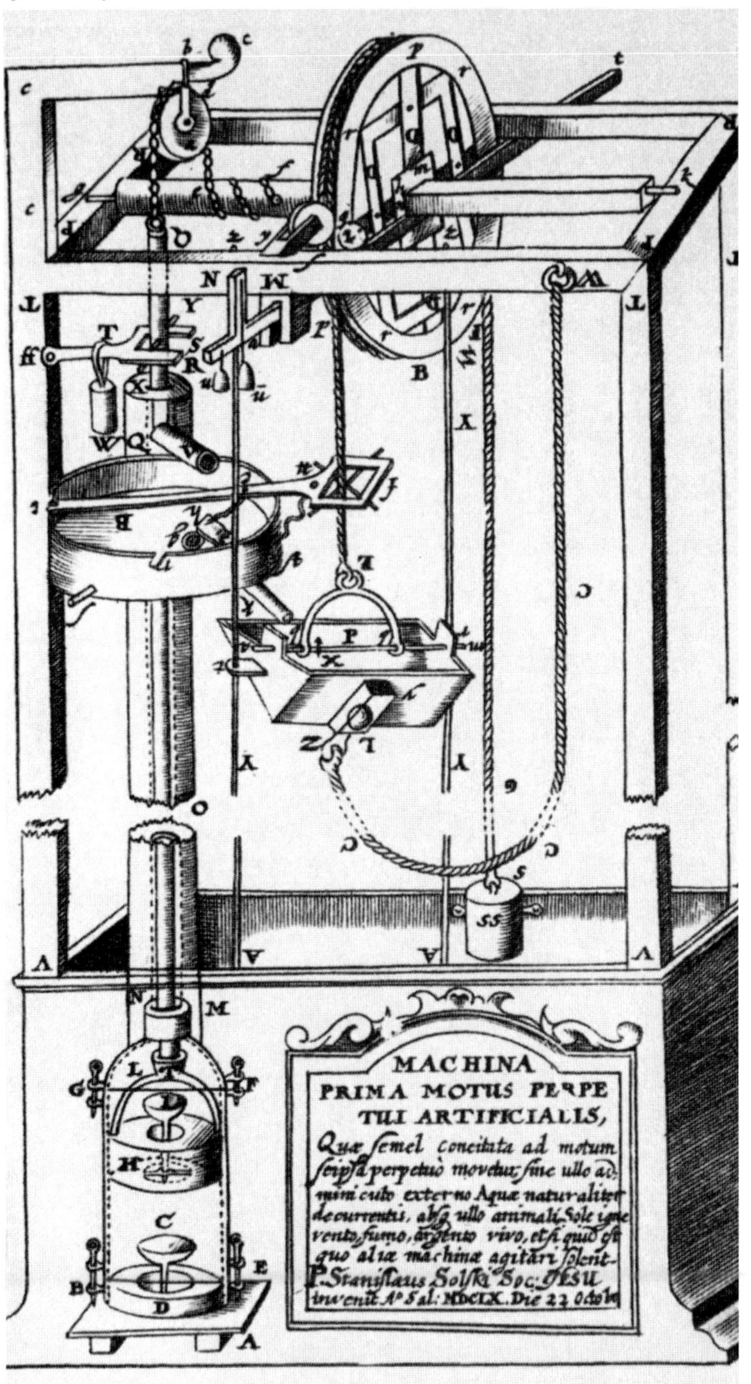

in heat, and T is the absolute temperature. Clausius used entropy to state the second law as:

The entropy of a closed system always increases.

Looking at this we see a problem. It applies only to closed systems, but in reality all systems are not "entirely" closed. No matter how well we try to isolate our system, there will always be some outside influences (for example, the Sun). Because of this, Clausius found he had to state the law as: *The total entropy of the universe is continually increasing.*

But what exactly is entropy? One way to describe it is to keep in mind a principle of the second law—that some of the heat extracted from a source is unavailable for conversion to work—Entropy change can be described as a measure of the extent to which heat is unavailable for conversion to work.

Disorder and Probability

The fact that the entropy of the universe always increases gives us some insight into this concept. Early on, most researchers in thermodynamics thought of heat as a fluid, but it was soon realized that this view was too restrictive and that it was more convenient to think of heat in terms of molecular motions. The molecules of a hot gas vibrate on the average more than those of a cold gas, but in both cases there is a range or distribution of velocities, some slow and some fast. Furthermore, when a hot gas comes in contact with a cool one, the molecules of the two gases collide, but on the average more vibrational energy is transferred from the hot gas molecules to the cool gas molecules than vice versa. Eventually, however, the vibrational energy is spread evenly throughout the gases, and equilibrium is reached. Since entropy is increasing in this situation, we can interpret it as a measure of the "evenness" in the way the energy is spread out. But this spreading out causes "disorder" in the system as the molecules intermingle, so an increase in entropy is also an increase in disorder.

A good way to illustrate the relation between entropy and disorder is to consider a jar containing marbles, half of them red and half of them blue, with all the red ones on the top. At this point our system has minimum entropy. If we begin

Above left: James Clerk Maxwell, who made important contributions to the kinetic theory of gases but is best known for his four equations that describe all of electricity and magnetism. Above right: Ludwig Boltzmann, an Austrian physicist who made important contributions to the kinetic theory of gases. Top left: The movement of the ball in a pinball machine illustrates chaos, or disorder.

to shake the jar we will see the marbles start to intermingle: A few red ones will start to move down, and some blue ones will move up. As we continue shaking, the marbles will become more and more mixed, and *disordered*. The entropy of this system is therefore increasing, and it is unlikely ever to decrease. For this to happen, the red and blue marbles would have to separate, and the probability of this is low.

Since entropy and disorder are synonymous, we can say that the "disorder" of the universe is also increasing, and never decreases. But the tendency of a system to pass from an orderly state to a disorderly state occurs because a disordered state is more "probable" than an ordered one; there simply are so many more disordered states. There is, in fact, a small probability that over a short interval of time the system will move to a more ordered state; but in the long run the disordered state will always prevail. This shows us that probability plays an important role in thermodynamics. This role was studied in detail by the Austrian physicist Ludwig Boltzmann (1844–1906). As a result of his work (and that of others), what is now a major branch of physics called *statistical mechanics* was developed.

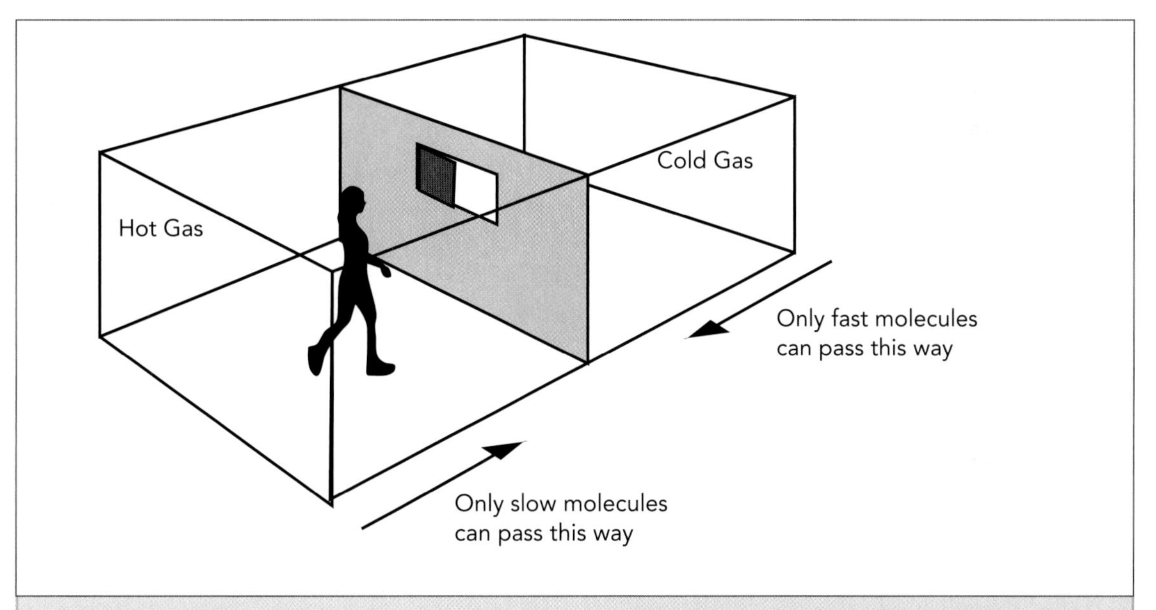

Maxwell's demon is a hypothetical being that knows the speed of all the molecules approaching a small window and can slide a door across the window at will to stop a particular molecule from getting through from either side.

MAXWELL'S DEMON

We saw above that random changes in a system occur with high probability, but not as a certainty. It is possible (but highly unlikely) that heat could flow from a cold object to a hot one. James Clerk Maxwell of Scotland showed this in an interesting way in 1867. He considered a system that was divided into two compartments. One side (A) contained a hot gas, and the other side (B) a cool gas; both gases had a distribution of molecular speeds. Maxwell visualized what is now called "Maxwell's demon," a superbeing that knew the speeds of each of the molecules. He was stationed at the partition between A and B, and this partition had a hole in it over which a small door could be slid. This creature would open the door when a slow-speed molecule from A was approaching it, and allow it to pass to B (the cool side). Furthermore, it would open the door when a high speed (hot) molecule from B approached it, and let it through. As a result, the hot side got hotter and the cool side cooler. Of course, the chances of this occurring in practice are infinitesimally small, but it does bring out an interesting point, and it shows the importance of probability in the second law of thermodynamics.

Third Law of Thermodynamics

The third law of thermodynamics was developed between 1906 and 1912 by the German physical chemist Walther Hermann Nernst (1864–1941). It can be stated as:

It is impossible by any procedure, no matter how idealized, to reduce any system to the absolute zero of temperature in a finite number of operations.

Alternatively, the law states that the entropy of a pure substance at absolute zero is zero, and therefore absolute zero is a reference point for determining entropy.

In 1911 the German physicist Max Planck (1858–1947) showed that the law was strictly true only for substances in a crystalline state, and this has been demonstrated experimentally. So a revised statement of the law is: *The entropy of a perfect crystal at absolute zero is zero.* This, of course, makes sense if we look at it. At absolute zero there is no thermal energy or heat, and therefore the molecules do not move. And if this is the case there can be no disorder.

Above: Walther Hermann Nernst, the German physical chemist who introduced the third law of thermodynamics. Top left: The third law of thermodynamics is said to be the definition of temperature.

ZEROTH LAW OF THERMODYNAMICS

The zeroth law was formulated after the first, second, and third laws, but is considered more fundamental, thus the use of *zeroth*, which precedes the number one, in its name. The zeroth law of thermodynamics is a relatively straightforward observation that deals with the equilibrium of systems. Consider two systems, A and B. If we place system A in contact with system B, and if after a certain length of time A and B have the same temperature, they are in equilibrium with each other. A logical test of this equilibrium is to use a third body, or test body. We can then state the third law as:

If two thermodynamic systems A and B are in thermal equilibrium, and B and C are also in thermal equilibrium, then A and C are in thermal equilibrium.

GIBBS AND FREE ENERGY

In a series of papers published from 1876 to 1878 in the *Transactions of the Connecticut Academy of Sciences*, Josiah Willard Gibbs (1839–1903) of Yale College applied the basic principles of thermodynamics. Previously they had been applied only to heat engines. His papers, which were titled "On the Equilibrium of

Heterogeneous Substances," are now considered to be one of the great achievements of nineteenth-century science. They are the basis of modern physics chemistry and of chemical thermodynamics.

Gibbs was interested in what caused chemical reactions to "go"—in other words, what was "driving" them. To explain it he introduced the concept of a chemical potential and free energy. They were the "forces" behind the reactions.

Among his other contributions was the formulation of the phase rule, which concerns the equilibrium between different phases of liquids, gases, and solids. He also laid the foundation for the science of statistical mechanics, and he invented vector analysis.

Despite all this it took several years for his work to be recognized, and still today most people are unfamiliar with his work. One of the main reasons for this is that it was published in an obscure journal; in addition, the rigorous mathematics used in the articles was difficult for most scientists to understand. One of the first to appreciate the importance of Gibbs's work was James Clerk Maxwell.

Josiah Willard Gibbs, an American physicist who applied the principles of thermodynamics to chemical reactions.

GIBBS'S PARADOX

As in the case of Maxwell's demon, we imagine a system divided in half by a partition. The two sides contain different gases at the same pressure and temperature. Removing the partition increases the entropy since there is an increase in disorder. Now, assume that the gases on the two sides are the same. When we remove the partition, nothing happens, and there is no increase in entropy. This seems to present a paradox, but Gibbs pointed out that it is important to take relevant information into account in such an experiment. The relevant information in this case is whether the gases are the same or not.

LIGHT AND OPTICS

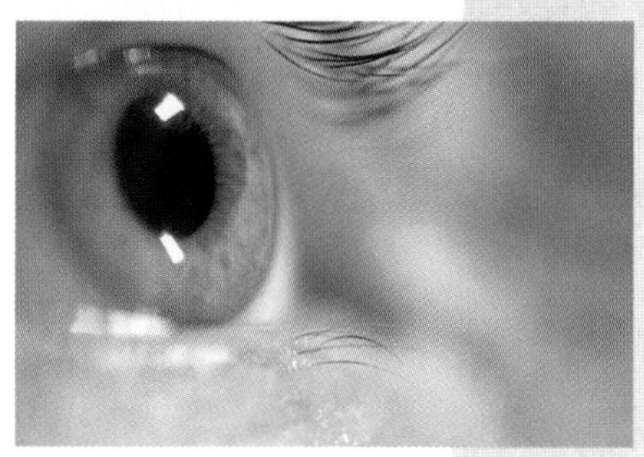

Left: Isaac Newton was the first to demonstrate that white light was composed of all the colors of the rainbow. Top: The Sun is our most important source of light. Light is also obtained from artificial sources, such as lamps. Bottom: The organ that allows us to detect light: our eyes.

Everything we have learned about the world is a result of our senses, and in many ways the most important of these is sight. Over the centuries we have learned a tremendous amount as a result of our sight. Our eyes, which are our organs of sight, are sensitive to light. Every thing we see is a result of the light that is reflected from or emitted by an object. The ultimate source of light is the Sun. It bathes the entire Earth and its inhabitants in light.

Many early scientists and thinkers were convinced that light had infinite speed, but Galileo decided to find out for himself and set up an experiment to do so. Using lanterns with shutters, he placed one at his side and told an assistant to carry one to the top of a mountain about a mile away. He instructed the assistant to uncover the shutter on his lamp as soon as he saw the light from Galileo's lamp. There was a small delay, but Galileo soon found that it did not vary with distance; it was just the reaction time of the observer. The only thing he could conclude was that the speed of light was extremely high, beyond the capacity of his equipment to measure. He was not able to determine if it was finite or infinite. And so began a series of investigations over hundreds of years to determine the properties of light.

Measuring the Speed of Light

Galileo did not determine the speed of light, but he played a critical role in its determination. In 1610, using a telescope he had constructed, Galileo discovered four tiny moons orbiting the planet Jupiter. They were particularly interesting in that they were periodically eclipsed by Jupiter. Within a short time their orbits were worked out so that their eclipse times could be predicted.

The Danish astronomer Ole Roemer (1644–1710) was observing these moons in the 1670s when he noticed that the eclipse times appeared to be inaccurate. At times they occurred several minutes earlier than predicted, and at other times they were late. He soon discovered that they were earlier when Jupiter and Earth were on the same side of the Sun and late when they were on opposite sides. Since it was known that Jupiter could be as close as 400 million miles from Earth when they were on the same side of the Sun, and as far away as 580 million miles when they were on opposite sides, Roemer came to the conclusion that the reason for the inaccuracy was that the light was taking several minutes to travel the extra 180 million miles when the two planets were on opposite sides. He used this to calculate the speed of light and got 150,000 miles/sec. It was considerably lower than the currently accepted value, but it was a good estimate, and it was the first proof that the speed was finite.

FIZEAU'S EXPERIMENT

Over a hundred years passed before anyone tried to come up with a more accurate value for the speed of light. French physicist Armand Fizeau (1819–96) set up an experiment in 1849 using a spinning disk with cogs and gaps along its outer edge; a mirror was positioned about five miles away. A light source was placed behind the spinning disk

Above: Armand Fizeau. Top left: Galileo Galilei discovered four of Jupiter's moons in 1610.

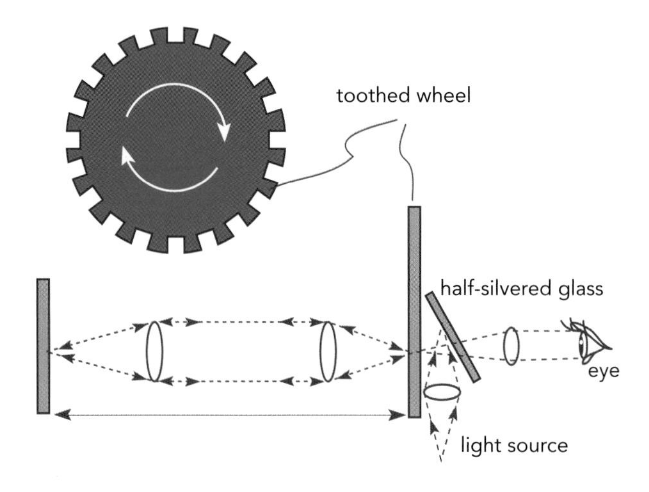

Fizeau's apparatus for determining the speed of light. Using a half-silvered mirror, he passed a beam of light through a cog in the wheel. It traveled to the mirror where it was reflected. If the speed of the wheel was exactly right, the pulse of light would return through the next cog.

so that a series of "light pulses" were sent to the mirror and reflected. The returning pulses were directed at the outer edge of the spinning disk.

It was obvious to Fizeau that the speed of light is so high that if the disk was spinning at a relatively slow speed, the pulse from a given gap would travel to the mirror and be reflected back before the gap could move out of the way. If the angular speed was increased, however, eventually a point would be reached where the cog adjacent to the gap would be in the way of the returning beam. In this case, an observer posted behind the spinning disk would not see the pulse. If the speed was increased further, eventually the pulse would start to pass through the next gap. As the speed increased further still, the intensity of the light would eventually return to a maximum when the next gap had moved so that the pulse passed entirely through it.

Using the known distance to the mirror and the angular speed of the disk, Fizeau was able to calculate the speed of light. His value was better than Roemer's but still only approximate. Fizeau then collaborated with Jean Bernard Léon Foucault (1819–68), also from France, in an experiment using a different and more accurate technique: In place of the cogged disk they used a spinning mirror. After obtaining a more accurate result, they decided to find out if the speed was different in water. Working separately, they both obtained the speed in water, but Foucault announced his results first (April 1850). He showed that water slowed the velocity by about 25 percent.

Albert Michelson worked tirelessly toward an accurate measurement of the speed of light. He won the Nobel Prize for Physics in 1909, the first American to do so.

MICHELSON AND MORLEY

One of the most accurate of the early determinations of the speed of light was obtained by Albert Michelson (1852–1931) of the University of Chicago and Edward Morley (1838–1923) of Case Western Reserve University in Cleveland, Ohio. The experiment was performed in the California mountains. The two men used a 22-mile path between two peaks, along with a special, eight-sided mirror that rotated. After obtaining a value in air, they decided to use a long vacuum tube so that they could measure the speed of light in vacuum. The beam was reflected several times and eventually traveled a distance of 10 miles. They obtained a value of 186,272 miles/sec in a vacuum, which was greater than in air.

CURRENTLY ACCEPTED VALUE FOR THE SPEED OF LIGHT

Over the next few years many other scientists measured the speed of light, so that we now know its value to a high degree of accuracy. The currently accepted value in a vacuum is 186,281.7 miles/sec, or in mks units, 299,792.8 km/sec. This is frequently approximated as 300,000 km/sec, or in cgs units as 3×10^{10} cm/sec. (The speed of light in a vacuum is usually designated as c.) Note that at this speed, light travels around the Earth seven times in one second and travels the 93 million miles from the Sun in eight minutes. Also, it is important to point out that the speed of light less in a transparent medium than it is in vacuum, although the difference for air is quite small.

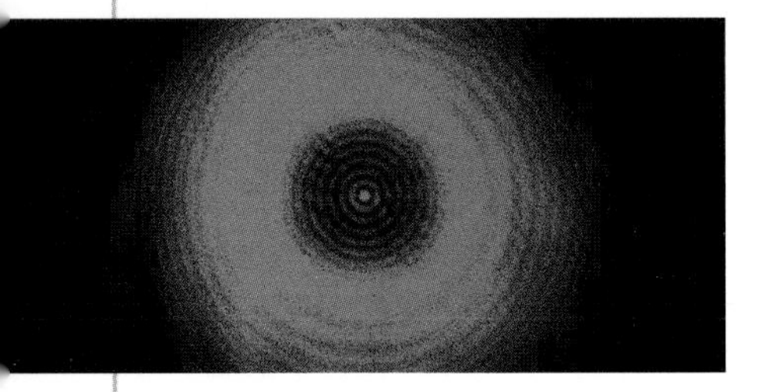

Wave or Particle?

What is light? That question was one of the major early problems in physics, and in a sense it is still a problem today. Even though we now know the velocity of light to a high degree of accuracy and have determined many of light's basic properties, we are still unsure of exactly what it is. The best we can do is describe how it acts. One of the earliest theories was put forward by Newton. He assumed that a light beam was made up of particles, which he referred to as "corpuscles." According to his theory, sources of light emitted corpuscles, and when they struck a surface they were reflected from it, and the reflected particles gave us an image of the object they were reflected from. Furthermore, these particles were stopped by opaque objects and passed through transparent ones. The theory appeared to explain many of the properties of light—for example, that light beams left sharp, distinct shadow edges and always traveled in straight lines.

About the same time, the Dutch scientist Christian Huygens (1629–95) put forward an entirely different theory: He proposed that light was a wave phenomenon. Futhermore, he introduced an interesting idea that is now known as Huygens's principle. It states that *each point of a wave-front of light may be considered a new source of wavelets.* Waves could, indeed, explain many of the known properties of light, but there were problems. Waves on water exhibited many well-known phenomena, and some of these phenomena appeared to be inconsistent with the properties of light. Nevertheless, the theory appeared to be an acceptable alternative to Newton's theory.

YOUNG'S EXPERIMENT

The impasse between the two theories of light remained for over a hundred years. No one was sure which was correct. Then in 1801 the English physicist Thomas Young (1773–1829) performed an experiment that demonstrated convincingly that the wave theory had to be the correct theory. He passed a beam of light through two side-by-side slits (see figure at right) and noticed that the emerging beams interfered with each other. At points where two crests came together they reinforced each other and the resultant waves had double the intensity of the individual waves. At points where a crest and a trough came together, they canceled each other and there was no light. These two cases are

Above: Thomas Young, the English physicist who proved the wave theory of light. Top left: The diffraction pattern created by a circular object is a series of light and dark rings caused by the interference of light waves.

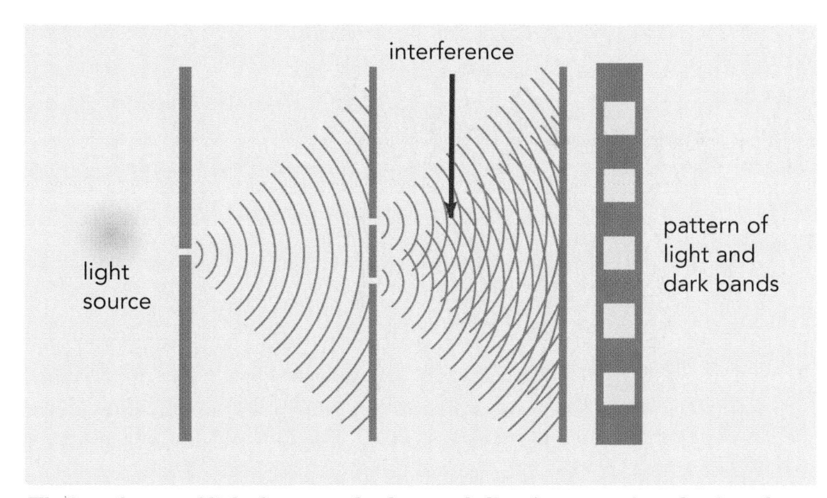

The interference of light from two closely spaced slits. At some points they interfere constructively and enhance the light; at other points they interfere destructively and decrease or eliminate the light. The result is a pattern of light and dark bands.

referred to as constructive and destructive interference respectively. The same phenomenon can be seen in water waves passing through two nearby slits. The experiment was a critical one in that it proved once and for all that light had wave properties.

Even though the evidence was clear, the wave theory was not accepted immediately. Over the next few decades, however, additional evidence was found, and the case for the wave theory was soon overwhelming.

DIFFRACTION AND POLARIZATION

Further evidence came in the form of *diffraction*, an interference phenomenon involving a single slit. In this case, light spreads into the region behind the slit. Actually, this also occurs at the edge of any object in a light beam under the proper conditions. Close examination of the shadow of any opaque object reveals that at its edge there is a series of dark and bright lines that extend outward from the boundary of the object. In particular, a close examination of the boundary shows that a small amount of light actually "bends" around the edge of the object. Diffraction can be explained using Huygens's principle.

POLARIZATION

Another phenomenon that verified the wave nature of light was *polarization*. To explain it, we first have to look at the nature of light waves. Earlier we saw that waves come in two types: transverse and longitudinal. In transverse waves, the wave motion is perpendicular to the direction of travel of the wave; in longitudinal waves it is parallel. After it was shown that light was a wave, it was soon shown that it was a transverse wave.

In effect, a light beam was made up of many "wavelets" vibrating perpendicular to the direction of its propagation. Furthermore, there is a material, called Polaroid, that allows only the light rays through that are vibrating in a particular direction; this direction is referred to as the axis of the Polaroid.

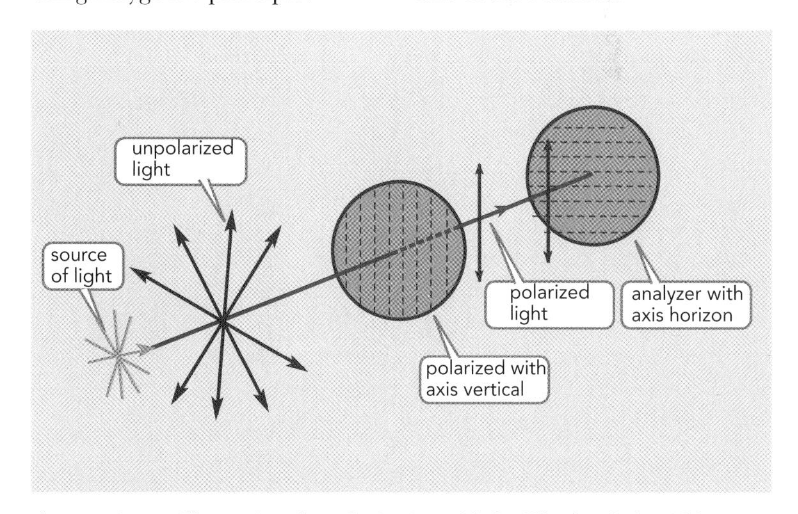

An experiment illustrating the polarization of light. The first Polaroid has its axis vertical, and only waves vibrating in this direction can pass through it. The second Polaroid has its axis set in the horizontal direction, and only waves vibrating in the horizontal direction can get through. As a result we see no light beyond the second Polaroid.

Light Basics

One of the most obvious properties of light is that sources of it vary in brightness. A 100-watt lightbulb, for example, is brighter than a 50-watt bulb. So we obviously need a measure of light *intensity*. The first measure to be used was a candle of specified size and composition; it was called the "standard candle" and had a brightness of one candle. Candles are, of course, not reliable sources, so the candle was eventually replaced by an electric bulb of a certain size; the unit was then called the "international candle."

If we step back from a light source it gets dimmer, so brightness also depends on our distance from the source. The reason for this dimming is that the light spreads out in a sphere around its source. As the sphere gets larger and larger, the amount of light striking a unit area decreases; and since we know from geometry that the area of a sphere is proportional to the square of its radius, we can say that the light intensity therefore drops off as the square of the distance. This means that if you double the distance, the intensity decreases by four.

Another unit of importance is the *foot-candle*. To define it we have to consider an imaginary spherical surface around the candle; if we assume it is one foot from the candle, then one square foot of this surface is illuminated by an intensity of one foot-candle. Furthermore, since light intensity is defined as the amount of light per unit area, it can be expressed in units of candles/ft^2. It is more convenient, however, to express this as *lumens*. The lumen is defined so that the intensity of one foot-candle has an intensity of one lumen/ft^2.

Above: Reflection is a light phenomenon that we experience in our everyday lives. Top left: Candlelight was the first measure used to define brightness, or intensity of light.

REFLECTION

Reflection is another well-known phenomenon associated with light. In dealing with it we have to consider two types of surfaces: flat and curved. From observation we know that when

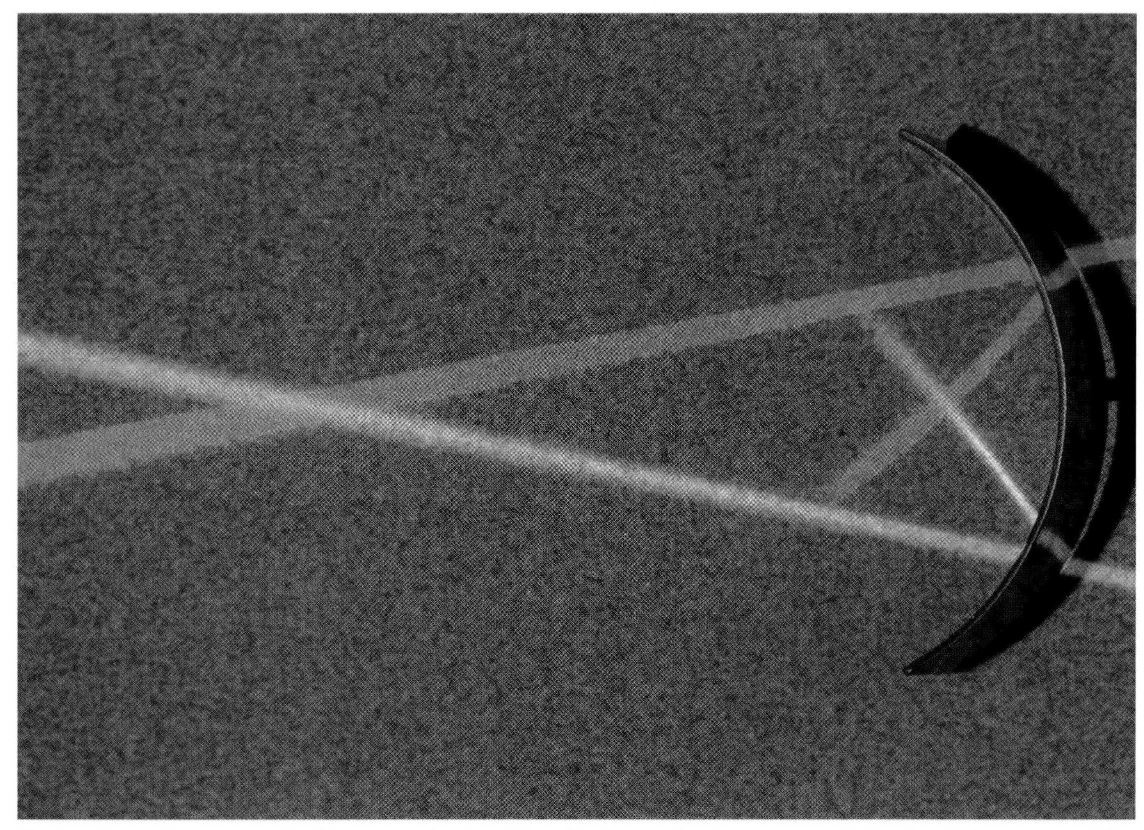

Beams of two colors are used to illustrate reflection. The reflecting surface is concave and therefore the beams cross. If the beams came in parallel they would cross at the focal length of the mirror.

a light beam is reflected from a flat surface—a plane mirror, for example—the reflected ray makes the same angle with the normal (perpendicular direction) as does the incident ray. Furthermore, the incident ray, the reflected ray, and the normal are all in the same plane, and the image is the same size as the object, and it is located the same distance behind the surface of the mirror as the object is in front of it. Finally, right and left in the image are interchanged as compared to the object.

Reflection from a curved mirror is a little more complicated. Two types of curvature are of interest: concave and convex. In most cases the curvature is assumed to be a section of a spherical surface. A mirror is concave if the reflecting surface is on the inside of the spherical surface; it is convex if the reflecting surface is on the outside. The amount that it curves is defined by the radius of curvature of the sphere.

Assume we have a concave mirror and allow rays coming in parallel to its axis to strike it. When the rays are reflected they will all focus at approximately the same point. This point is called the *focus*.

The view from the sideview mirror of a car. This mirror is usually curved to make it wide-angle, which makes the cars appear to be farther away than they actually are.

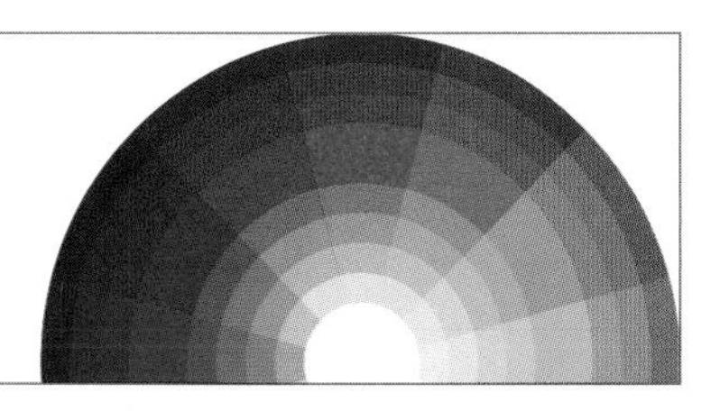

More Properties of Light

Light is a wave and therefore has all the properties of waves that we described earlier. One of the most important of these properties is wavelength. Thomas Young was the first to calculate the wavelength of white light. Using data from the interference fringes in his double-slit experiment, he found it to be about 50 thousandths of an inch. Today we usually express this wavelength in millimicrons (mμ) where 1 mμ is a billionth of a meter (also referred to as a nanometer). Another unit that is commonly used is the angstrom (Å), named in honor of Anders Ångström (1814–74), who originated it; it is one-tenth of a millimicron. In a later section we will see that the wavelength of light actually depends on its color. Red light, for example, has a wavelength of 760 mμ, or 7,600 Å, longer than that of blue light. In angstroms, the wavelength of all the colors of light ranges from 7,600 for red to 3,800 for violet.

Since light has a wavelength it also has a frequency and it has amplitude, which is a measure of the brightness or intensity of the source.

Isaac Newton observing the spectrum of a beam of white light passing through a prism. Top left: A color wheel illustrates the color spectrum.

The spectrum of colors includes all the colors of a rainbow.

DISPERSION

In 1666 Newton allowed a beam of white light to fall on a prism and noticed that the prism spread it out into all the colors of the rainbow. The effect had actually been noticed earlier, but Newton was the first to explain where the colors came from. He placed a second prism next to the first one and showed that the spectrum of colors could be combined back into white light. On the basis of this he postulated that white light was made up of light of all colors. In effect, when the white light beam entered the prism it was dispersed into all the colors of the rainbow. These colors are (from longest wavelength to shortest) red, orange, yellow, green, blue, and violet. They are now referred to as the spectrum of colors.

Another breakthrough was made in 1801 when Thomas Young showed that all the above colors could be obtained from just three colors—red, green, and blue—mixed in appropriate combinations. They are referred to as *primary colors*. When all three of these primaries are mixed together they give white. Furthermore, when two colors mix together to give white, they are referred to as *complementary*. An example is yellow and blue. Indeed, any color toward one end of the color spectrum has a complementary color near the other end.

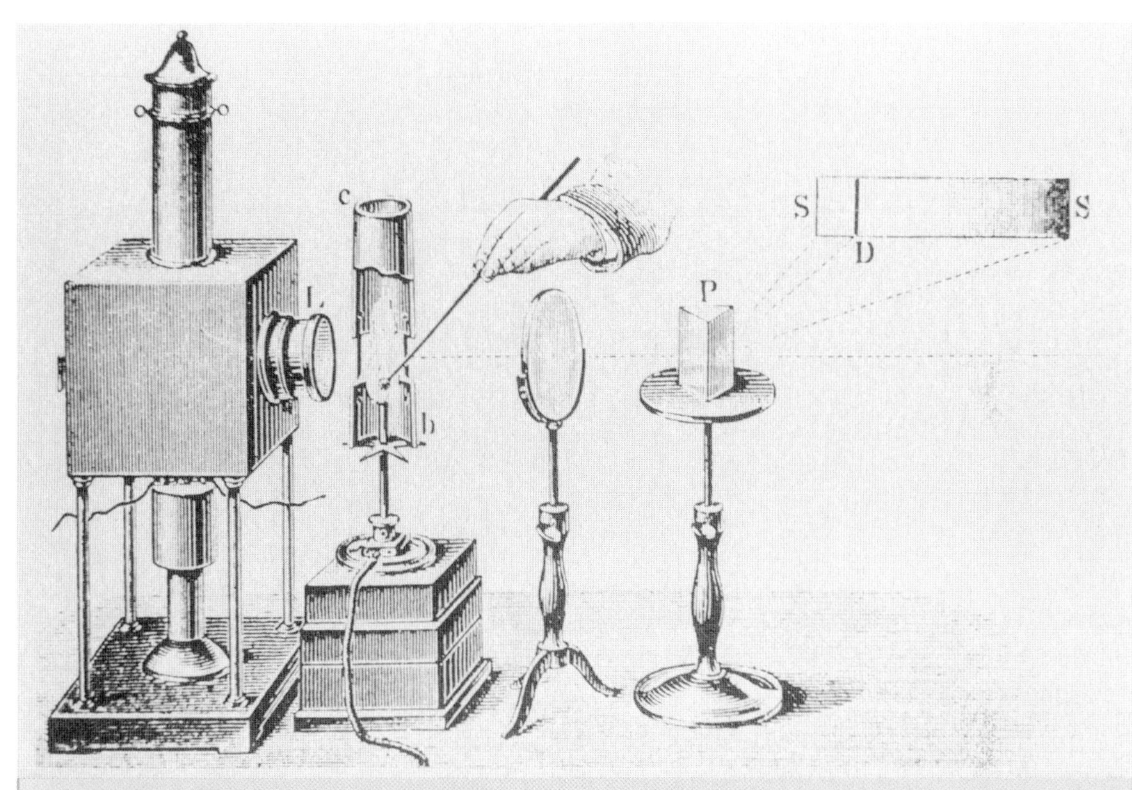

An early spectroscope, dating from about 1900.

THE VISIBLE SPECTRUM

The English physicist William Wollaston (1766–1828) passed a beam of light from the Sun through a prism in 1802 and noticed that a number of dark lines were superimposed on the color spectrum. The German physicist Joseph von Fraunhofer (1787–1826) later found over 600 of these lines, determined their wavelength and cataloged them; these are now referred to as Fraunhofer lines.

Then in 1859 the German chemists Gustav Kirchhoff (1824–87) and Robert Bunsen (1811–99) developed the spectroscope. It consisted of a slit through which light enters, collimating lenses, which gather light together in a parallel beam, a prism, and a screen for observing the spectrum. It allowed them to examine the spectra of various materials in detail, and they were able to show that each substance had its own unique set of dark lines. Furthermore, they found there were three types of spectra: continuous, bright line, and dark line. Continuous spectra were given off by a white light (from heated, white-hot metal, for example); bright-line spectra were produced by a heated, luminous gas; and dark-line spectra were produced when white light passed through a cool gas. Kirchhoff showed that there was a relationship between bright and dark lines in spectra. Hot, glowing sodium, for example, produced a number of bright lines. And if white light (with its continuous spectrum) was passed through cool gaseous sodium, it produced dark lines in the same regions of the spectrum.

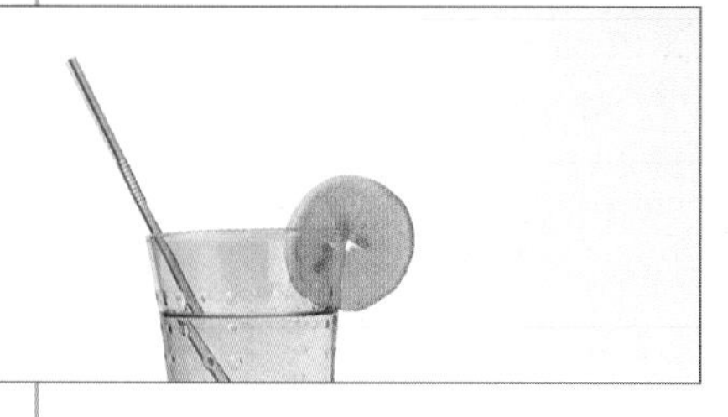

Refraction

If you have ever placed a stick or similar object in water, you know that it appears to bend at the point it enters the water. To see what causes this, consider a light ray that passes through a piece of glass with parallel sides. If the ray enters the glass perpendicularly (at an angle of 90 degrees) it will pass through the glass undeflected; but if it comes in at an angle, it will be bent. To see why, consider the wave fronts approaching the glass (see figure). The bottom side of the wave front enters the glass first, and because light travels slower in glass than in air, this part of the wave front slows down. From the diagram we see that it therefore lags behind the upper section, and this causes the overall ray to bend. When it exits the glass, however, it bends back so that the exiting beam is parallel to the incident beam. This happens when a light beam goes from a less-dense to a more-dense transparent medium (or vice versa), and it is called *refraction*.

Referring to the figure, we call the angle between the normal and the incident ray the incident angle, i, and the angle between the normal and the refracted ray the refracted angle, r. If the beam is entering a denser medium, it will be bent toward the normal. If it enters a less-dense medium, it will be bent away from the normal. Note also that the incident ray, the refracted ray, and the normal are all in the same plane.

INDEX OF REFRACTION

As we saw above, refraction occurs because the speed of light is different in different materials. We can define the relative index of refraction n_r as the ratio of the two velocities. Furthermore, if we assume one of the media is empty space (where the velocity of light is c), we get the *absolute index of refraction*, where $n = c/v$, and v is the velocity of light in the medium. The index of refraction of water is 1.33. For glass, it varies between 1.5 and 2, depending on the type of glass.

Knowing the indices of refraction of two media, and the incident angle of the light beam, we can calculate the

incident ray
normal
air
water
refracted ray

This diagram illustrates a beam of light in air passing into water. Because the index of refraction of water is greater than that of air, the beam is refracted, or bent, toward the normal (dotted line). Top left: Refraction can be observed simply by placing a straw in a glass of water.

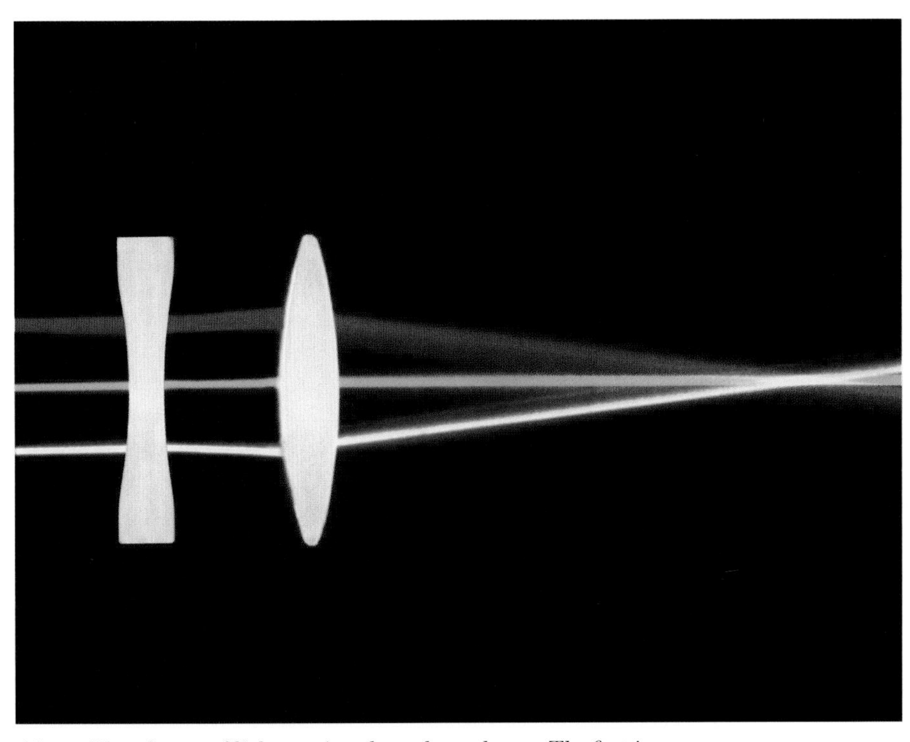

Above: Three beams of light passing through two lenses. The first is a concave lens, the second a convex lens. The concave lens diverges the rays, the convex lens causes them to converge. Below: When a beam of white light strikes a prism, the light breaks up into a rainbow of colors.

refracted angle. The law giving this angle was first worked out by Dutch astronomer and mathematician Willebrord Snell (1580–1626) in 1621 and is now known as Snell's law. It is given by

$$n_1 \sin i = n_2 \sin r$$

where n_1 and n_2 are the two indices of refraction.

TOTAL INTERNAL REFLECTION

So far we have considered a beam that enters a denser medium. What happens if the beam goes from a denser to a less-dense medium, such as from water to air? According to the rule in the previous section, the beam will bend away from the normal. Furthermore, as we increase the angle of incidence, the refracted ray will bend even more. Eventually, for a particular angle of incidence, the emerging beam will just graze the surface. In this case the refracted angle will be 90 degrees. The incident angle where this occurs is called the *critical angle.* Beyond this angle something strange happens, and we find that the light is totally reflected. In the case of water and air, the critical angle is 48.6 degrees; for glass and air it is approximately 42 degrees. If you looked at the surface from beyond the critical angle it would look like a mirror. Because prisms can act as mirrors, they are very useful in optical instruments. Note that for a ray approaching the surface of a prism perpendicularly, you have total internal reflection, causing the diagonal to act as a mirror. This is why prisms are used frequently as mirrors in telescopes, binoculars, and cameras. Prisms reflect 100 percent of the light incident on them; no other reflecting surface is this efficient.

Geometrical Optics

We know that a ray of light bends when it hits a flat piece of glass at an angle. But what happens to a ray when it enters a lens? Assume the lens has a double convex surface similar to that of a hand magnifier. If we examine the rays' striking points along the lens, we find that a ray entering near the edge of the lens will be bent the most. Rays closer to the center will be bent less, until finally at the center they will pass through undeflected.

If the rays approaching the lens are all parallel to the axis of the lens they will all converge to the same point. This point is the *focal length* of the lens. Any image that is found here is referred to as *real* in that it can be viewed on a screen. Furthermore, just as we have both convex and concave mirrors, so too do we have convex and concave lenses. In a concave lens, if we again assume the rays come in parallel, they will diverge as they pass through the lens. We can still determine the focal length of the lens, however. We merely have to trace the refracted rays back through the lens. In this case the dotted lines will come together on the left side (the side from which parallel rays are approaching), and the distance from this point to the center of the lens is the focal length. If an image is found here it is referred to as *virtual*. It cannot be viewed on a screen, but if you look back through the lens you can see it.

The lenses described above were double convex and double concave, but other combinations are possible. One side can

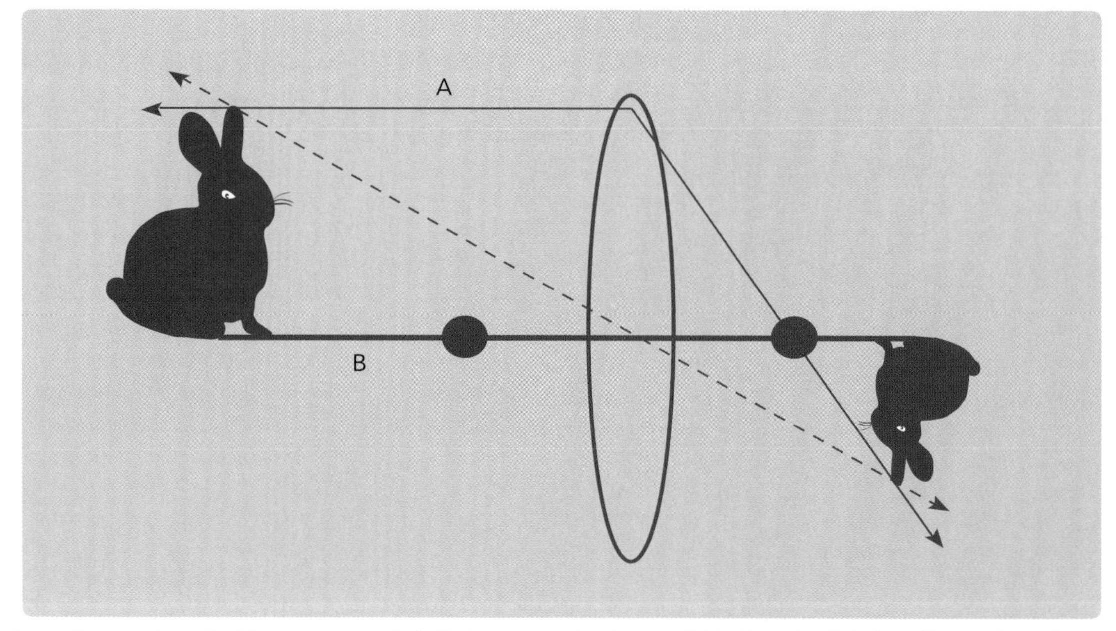

A ray diagram for a double convex lens. Note that a ray coming in parallel to the axis of the lens (A) is refracted so it passes through the focal point on the other side of the lens. A ray passing through the center of the lens (B) is undeflected. The image is formed where the two rays cross. Top left: A magnifying glass utilizes a double convex lens.

be plane or flat; such a lens is referred to as plano-convex (or plano-concave). In addition, both sides could be shaped the same way but with different radii of curvature; they are referred to as meniscus lenses.

RAY DIAGRAMS FOR LENSES

Assume we have an object a certain distance in front of a double convex lens. Its distance from the lens is called the *object distance.* An image of the object is formed a certain distance behind the lens. The distance in this case is referred to as the *image distance.* If we know the object distance and the focal length of the lens, we can determine the image distance; we can also determine how large the image is, whether it is right side up or inverted, and whether it is real or virtual. One method of doing this is called the ray diagram. We will illustrate the technique using a double convex lens, but it also applies to any type of lens. In this method you begin by drawing a line from the top of the object parallel to the axis to the lens, then to the focal point on the other side of the lens. A second line is drawn from the top of the object through the center of the lens and straight on beyond the lens. The point where the two lines cross is the tip of the image; the base of the image is at the center line. If the object is closer to the lens than the focal point, the image will be on the same side as the object. In this case you will have to dot-back on the rays to determine the position of the image.

A large astronomical telescope. Most telescopes used for studying the skies are reflectors.

TELESCOPES

The first practical telescope was invented by the Dutch optician Hans Lippershey (1570–1619) about 1604, but he kept it secret for several years. About five years later, however, Galileo heard of it and constructed one for himself.

Telescopes are generally of two types: refractors and reflectors. The main component of a refractor telescope is a convex lens, which is referred to as the *objective.* Light passes through this lens, forming an image that is further magnified using an eyepiece. The objective of a reflecting telescope is a concave mirror. Again, the image is viewed using a small eyepiece.

In the case of both reflectors and refractors, the bigger the objective, the greater the "light-gathering power" of the telescope. High light-gathering power is important because it allows the image to be magnified without "weakening" it (making it appear washed out). Also important is the telescope's *resolving power*, which is determined by the size of the objective and also by the quality of the optics throughout the telescope. This is the ability to resolve or separate two closely associated objects, such as two stars.

Because our atmosphere hinders our view of objects in space, many telescopes have been placed into orbit. One of the largest and most expensive is the Hubble telescope, which was launched in 1990. It has given us an unprecedented view of the heavens.

CHAPTER 7

ELECTRICITY AND MAGNETISM

Electricity and magnetism were both known to early man. There are, in fact, indications that magnetism was known as early as the Iron Age. No one is sure where the term "magnet" came from. According to one legend, it came from the name of the land, Magnesia, where lodestone was first found. According to another it came from a shepherd named Magnes, who was bewildered when he found that his shoes, or more exactly, the nails in his shoes, were attracted to lodestone in the ground. The word "electricity," on the other hand, almost certainly came from the early Greek word *elektron*, which means amber. Amber was one of the first materials known to accumulate a charge, although at that time it was not thought of as a charge. When rubbed it attracted lightweight objects such as lint or straw, but no one had any idea why. It was soon discovered that charges sometimes attracted and sometimes repelled one another, just as "magnets" from lodestone did. The electrical force, however, was much weaker than the magnetic force.

At first, both phenomena were regarded with little more than curiosity. They were interesting, but appeared to be of little use to man.

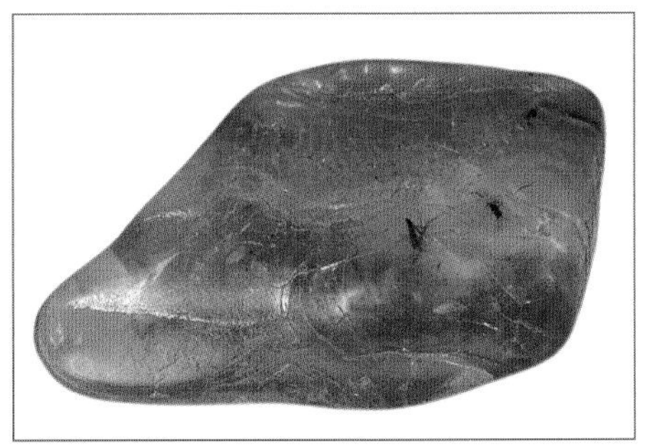

Left: A lightning bolt. When the potential difference between two clouds gets high enough, a discharge in the form of a huge electric current takes place between them. Top: This boy's hair has been charged with static electricity, and because like charges repel, the individual hairs on his head repel one another, and therefore stand out. Bottom: A piece of amber. Amber played an important role in the history of electricity because it acquires an electric charge when rubbed.

Static Electricity

Most early electrical experiments used what is called the Leyden jar, so named because it was invented at Leyden University in Holland. It was a large glass jar with tinfoil inside and out, a wooden lid, and a brass rod that extended down through the lid and was attached to copper wire that was in contact with the inner foil. It had the property of being able to store electrical charge, which at that time was thought of as an "electrical fluid." It was, in effect, a capacitor, and many important experiments were performed using it. Some of the first were performed by the English experimenter Stephen Gray (1666–1736). In 1729 he noticed that when a long glass tube was "electrified" by rubbing it, the corks at the end of the tube were also electrified, indicating that the electrical fluid had moved. He continued experimenting with other materials and soon found that the electrical fluid moved much more easily in some materials than in others. As a result, he classified some materials as "conductors" and others as "insulators." The electrical fluid passed easily through conductors but not insulators. Gray also discovered that the electrical fluid was on the surface of objects and not in the interior.

Above: The Leyden jar played a central role in early physics. It could be used to store electrical charge, so it was a simple form of a capacitor. Top left: An electrical discharge to a finger.

At about the same time, a French physicist, Charles du Fay (1698–1739), discovered that two electrified objects sometimes attracted each other and sometimes repelled each other. In particular, if both were electrified in the same way, they repelled, but in some cases, if they were electrified in different ways, they would attract. On the basis of this, he postulated the existence of two types of electrical fluid, which he referred to as vitreous electricity and resinous electricity. They eventually became known as positive and negative electricity.

FRANKLIN'S KITE

The next advances were made in America by the statesman and scientist Benjamin Franklin (1706–90). Franklin's interest in electricity was kindled in 1746 when he attended a public lecture in Boston. After the lecture he purchased a Leyden jar and began experimenting. It was well known at the time that if the jar was charged and you touched it, you would get a shock. Furthermore, a fairly large spark would jump from the ball at the top (which was connected to the inner foil) if a wire from the outer foil was brought near it. Franklin noticed the similarity between this spark and lightning, and he wondered if they were one and the same. To satisfy his curiosity he flew a kite during a thunderstorm, equipping it with a pointed wire that he connected to a silk thread. He tied a metal key to the other end of the thread. During the storm Franklin put his hand near the key, and it sparked just as a charged Leyden jar did. He also used the key to charge a Leyden jar.

AMBER AND GLASS

How was the electrical charge that was stored in a Leyden jar usually generated? For an answer to this, we have to go back to the early Greeks. They discovered that an amber rod, if rubbed with fur, would attract straw or feathers or other lightweight objects. We now know that it accumulated a negative charge. Similarly, when a glass rod was rubbed, it accumulated a positive charge. It was this charge that was transferred to the Leyden jar. Franklin, however, did not see this charge as two fluids; he believed that when the amber rod was rubbed with fur, a negative charge was deposited on the rod from the fur. In the case of the rubbed glass rod, he believed the negative charge was drawn out of the rod, and since it began with equal amounts of positive and negative charge, it was now left with positive charge. In essence, only one type of charge flowed.

Benjamin Franklin performing his famous kite experiment. A metal wire on the kite attracted a lightning strike, and electricity flowed down the wire to a key, which was used to charge a Leyden jar. The experiment proved lightning was an electrical phenomenon.

The experiment convinced Franklin that the electrical fluid of the Leyden jar was also present in the clouds during the storm. And in performing it he was extremely lucky: The next two men that tried the experiment were both killed by electrical discharges. Franklin's experiments showed that lightning was attracted to sharp objects, and as a result pointed rods, or "lightning rods," were soon placed on the roofs of houses to protect them. If the rod was connected by a wire to the ground, the electricity in overhead clouds was discharged through the rod to the ground. Franklin also decided, as a result of his experiments, that Du Fay's two-fluid theory of electricity did not fit the facts well. He suggested a one-fluid theory.

Lightning rods safely dissipate electrical charges so that buildings and other structures are not damaged.

Electrostatics

Once it was established that like charges repelled each other and unlike charges attracted each other, the question was, how strong was the force between them? This was taken up by the French engineer Charles Coulomb (1736–1806) in 1784. Using a very sensitive device he had just invented called a torsion balance, he was able to show that the force of electrical attraction or repulsion between two charges was proportional to the product of the two charges and inversely proportional to the square of the distance between them. It was, in effect, the same as the gravitational force between two masses (in this case, however, charge took the place of mass). It became known as Coulomb's law.

To use the law, we need to know the units of electrical charge. The common mks unit is the coulomb.

ELECTRIC FIELDS

According to Coulomb's law, two charges attract or repel each other with a certain force, according to the type of charges. But there is another way of looking at this, and it was favored by the English physicist Michael Faraday (1791–1867). He introduced the idea of "lines of force"

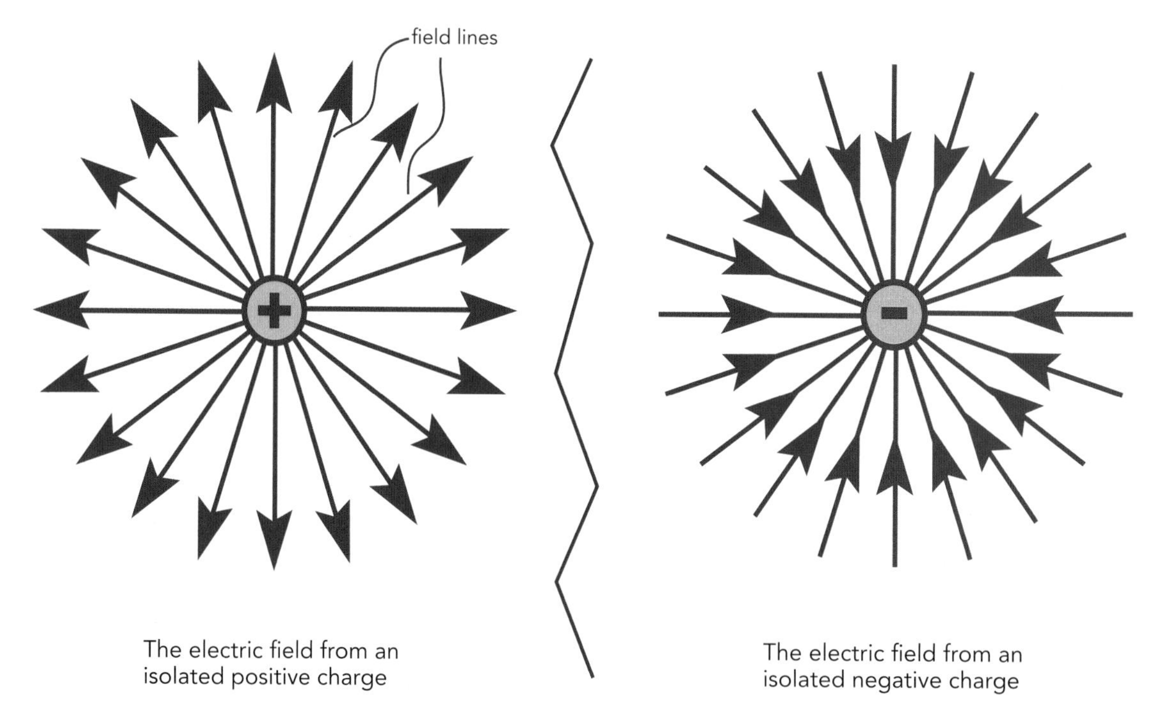

Above: The field lines of a positive electric charge are directed outward from the charge (left) and those of a negative electric charge are directed toward the charge. Top left: A drop of nickel-zirconium, heated to luminescence, appears to float in the region between two electrically charged plates. It is held in position by an electric field that counters its gravitational pull downward. This process is referred to as electrostatic levitation.

in 1844, and although it was not taken seriously at first, it was the beginning of the idea of "fields." According to this idea, every charge is surrounded by a field that depends on the strength of the charge (the greater the charge, the greater the field). If a second charge is brought into a field, it feels the first charge and acts accordingly. The direction of the electric field is such that like electric charges will repel and unlike attract. In effect, the lines of force from a positive charge are directed outward from the charge, while those from a negative charge are directed inward, toward the charge. The spacing of the lines gives a measure of its strength.

POTENTIAL DIFFERENCE

When you take a charge and move it toward another charge of the same type, there is a repulsion between the two charges, and therefore it takes work to perform the task. You have the same thing in the case of Earth's

Charles Augustin Coulomb, the French physicist who first measured the force between electric charges.

gravitational field. When you lift a box a certain distance upward against Earth's gravitational pull, you do work. At the same time you give the box potential energy. In the same way, if you move a charge between two different points in an electric field, you change the potential energy. We say that the two points have a potential difference, and we can define the potential difference V between two points in an electric field as the work done moving a unit charge from one point to the other. In mks units the unit of potential difference is the volt. It is formally defined as the potential difference between two points in an electric field such that one joule of work is done in moving a charge of one coulomb between the points.

In addition to defining the potential difference between two points, we can also define the absolute potential at a point. This is the potential difference between a point and an arbitrarily selected zero of potential. The zero is usually taken as the potential of ground. When a body with charge is connected to the earth via a conductor, the charge will flow until the potential difference between the body and the earth is zero.

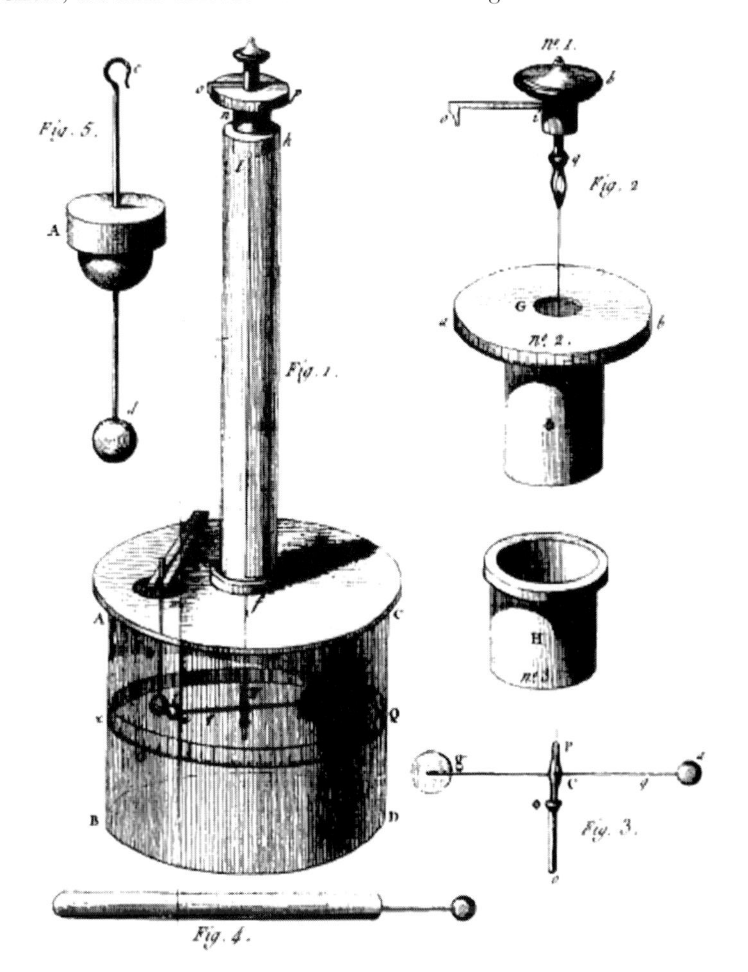

A torsion balance similar to the one used by Coulomb to measure the force between electric charges.

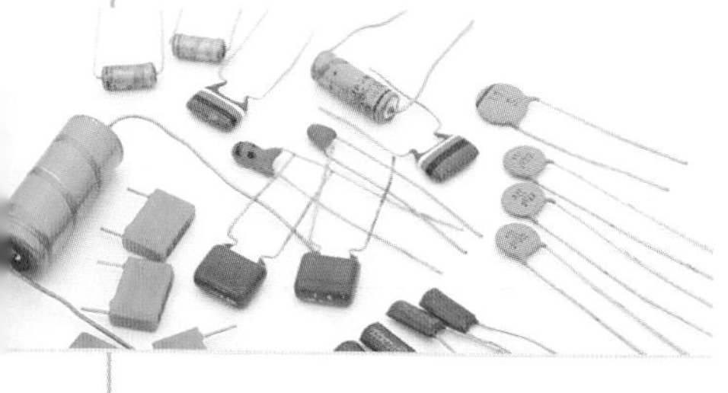

Capacitors

One place where we find an electric field is in a capacitor. Capacitors are now used commonly in most electrical applications, particularly in electronic devices. As we saw earlier, the Leyden jar was a capacitor. The easiest way to visualize a capacitor is to consider a conducting metal plate. Assume we keep it insulated and place a negative charge on it, then bring a second conducting plate that is positively charged up to it, making sure the two plates do not touch. The charge on each of the plates will be attracted to that on the other. The result will be an electric field between the plates, with the direction of the field from the positive to the negative charges.

If we assume there is a vacuum between the two plates, we can assume that the charge (call it Q) on the plates will remain constant, and also that the potential difference between the plates will be constant. We define the ratio of them as the capacitance C, such that

$$C = Q/V$$

Above: Teflon is frequently used as a dielectric in capacitors. Top left: A selection of modern capacitors.

Charge +Q
-Q Charge
Plate area A
Electric field E
Plate separation d

A positive electric charge accumulates on one side of this capacitor, and an equal negative charge on the opposite side. In the region between them there is an electric field; its strength depends on the amount of charge on the plates.

The spark plug in this lawn mower is fired by a capacitor discharge.

where Q is in coulombs, V is in volts. The unit of capacitance is therefore coulombs per volt, which is called the farad, in honor of Michael Faraday. As it turns out, the farad is a rather large unit, and it is more common to see capacitors with capacitances of microfarads (μf), which is a millionth of a farad.

In practice, the space between the plates is usually not a vacuum but is filled with a material called a dielectric. A dielectric increases the capacitance, and a measure of this increase is the dielectric coefficient. The dielectric coefficient of mica, for example, is approximately 4; this means that if mica is placed between the two plates, the capacitance is increased by 4 (compared to vacuum). Other dielectric coefficients are: 5 for glass, 3 for rubber, 3.4 for Plexiglas, and 2 for Teflon. In addition, capacitors are usually made up of several plates rather than just two, and the plates are frequently rolled up into cylinders.

A girl with her hand on a Van de Graaff generator. It produces an electric charge that causes her hair to stand out because each strand has the same charge and like charges repel.

VAN DE GRAAFF GENERATOR

An interesting device for accumulating a large amount of charge was designed by the American physicist Robert Van de Graaff (1901–67) in 1931. In his device a large hollow metal sphere is set on top of an insulating tube. Inside the tube is a belt driven by a small motor. A series of needles, which are kept at a high negative potential, touch the belt at the lower end. Another series of needles at the upper end transfers the charge to the sphere. A third set of needles creates a high potential, which ionizes the air in the region inside the sphere. As a result, positive charges are attracted to the belt and carried downward. The net result is a huge buildup of negative charge on the sphere.

In many demonstrations, two such models are used, with one collecting a large amount of negative charge and the other a large amount of positive charge. Eventually, when they accumulate enough charge, a "lightning bolt" passes between the two spheres.

Magnetism

Records indicate that the early Greek philosopher Thales studied magnetism about 6000 BCE and that the Chinese used simple compasses composed of "lodestone" as early as 2000 BCE. Lodestone is a naturally occurring iron ore that is found in many places on Earth; it is usually referred to as magnetite.

It was soon discovered that the ends on a sliver of lodestone always pointed toward the Earth's North and South poles. Magnets were soon said to have two poles: a north pole (N) and a south pole (S). Also, it was discovered that like poles (N-N or S-S) repelled each other, and unlike poles (N-S) attracted each other, much in the same way electrical charges did. By the early 1200s, slivers of lodestone were being used as compasses on board ships.

Above: A small piece of magnetite, which is also known as lodestone. It is a magnetic material that is an iron oxide. Magnetite is the most magnetic of all the magnetic minerals on Earth. Top left: A compass.

About 1600, William Gilbert (1544–1603), physician to Queen Elizabeth I, suggested that, because a compass needle always pointed to the North and South poles of the Earth, the Earth had to be a huge spherical magnet. He published a book titled *De Magnete,* in which he discussed the idea and summarized the current knowledge of magnetism.

In 1785 Charles Coulomb measured the force between magnetic poles at various separations using his torsion balance. He found that the force was similar to that between electric charges; in other words, it was proportional to the product of the strength of the poles, and inversely proportional to the square of their distance of separation. This applied to both attracting and repelling poles.

MAGNETIC FIELDS

Much as an electric charge is surrounded by an electric field, a magnetic field surrounds the poles of a magnet. One of the easiest ways to see what this field looks like is to place a sheet of paper over a magnet and spread iron filings on it. The filings take

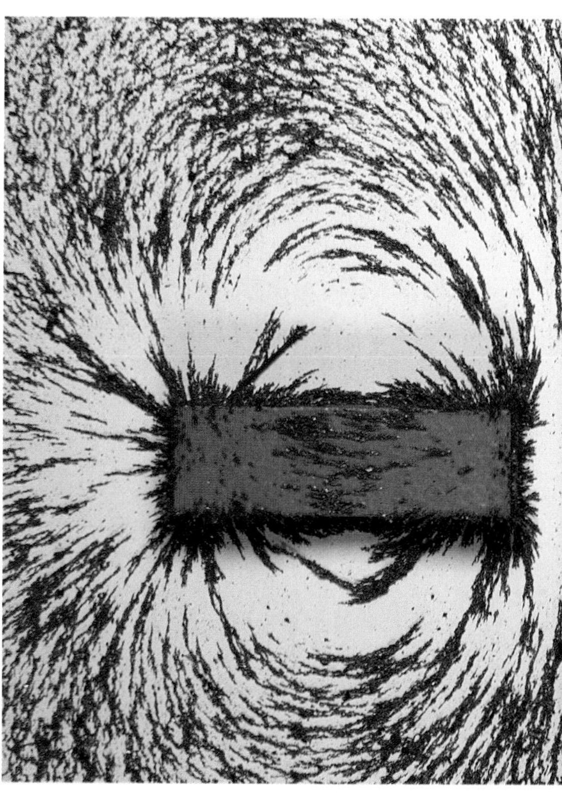

Iron filings have been placed over two magnets to show the field lines of the magnets. In this case the facing poles have the same polarity and therefore repel one another.

Several pieces of cobalt. Cobalt is a magnetic element that is used in making magnets, various alloys, electroplating, and for battery electrodes.

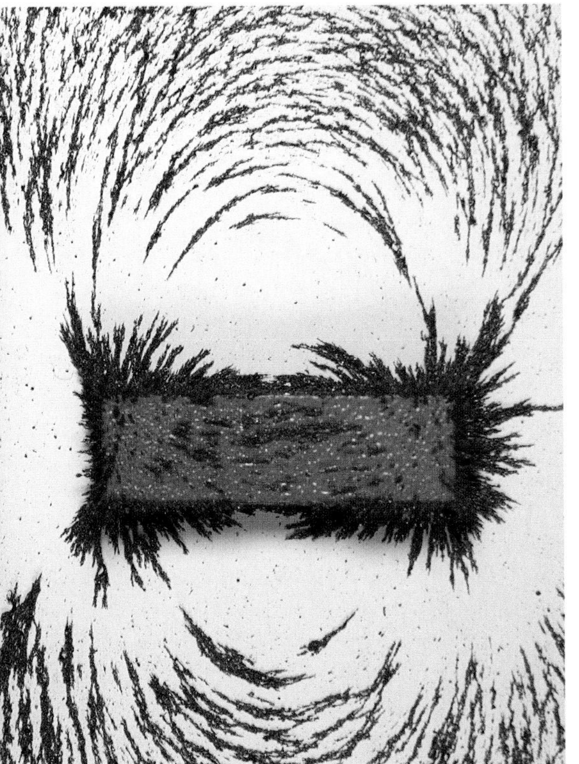

up a definite pattern: They create lines that curve from one pole of the magnet to the other. The lines are crowded together near the poles and spread out as they move away from the poles.

The unit of magnetic force in mks units is the *weber*; it is actually a unit of magnetic flux, which is the number of lines of magnetic force passing through an area perpendicular to the lines. More useful is the magnetic flux density, which is the number of lines passing through a unit area. In mks units this unit area is the square meter, so the units of magnetic flux density are webers/m^2. In cgs units the unit area is a square centimeter, and the flux density is known as a gauss.

OTHER PROPERTIES OF MAGNETS

Iron is the metal we usually associate with magnetism, but several other metals are also attracted to magnets, including steel, cobalt, nickel, and gadolinium; they are referred to as ferromagnetic materials. One of the first things we learn about magnets is that you can generate another magnet by stroking a piece of iron with a magnet. Iron magnets, however, can easily be demagnetized and are therefore referred to as temporary magnets. Steel, on the other hand, is harder to magnetize than iron, but once magnetized it is hard to demagnetize. It is therefore referred to as a permanent magnet.

Why are ferromagnetic materials magnetic, and other metals not? The answer centers on what are called magnetic domains. If you examine the structure of a ferromagnetic material with a powerful microscope, you see that small regions of "tiny magnets" exist throughout. In natural iron these regions, or domains, are not lined up, so their north and south poles are randomly oriented, and the overall material is not magnetic. If you bring a magnet nearby and stroke it along the material, however, it lines up the domains and creates a magnet. Tiny magnets such as this exist in all materials, but magnetic domains exist only in ferromagnetic materials.

Currents

Soon after the invention of the Leyden jar, the discovery was made that electric charges could move from one point to another. In particular, if you touched the knob at the top of the jar you got a shock. But this was only a temporary movement of charge; it stopped immediately upon the discharge of the Leyden jar. Scientists wondered if it was possible to create a continuous transfer of charge—but it was obvious that a source of some sort would be needed to produce it.

Luigi Galvani.

GALVANI AND THE FROG EXPERIMENTS

It might seem strange that frogs played an important role in the discovery that led to a continuous transfer of charge, but they did. The first step toward the discovery came in 1782 when Luigi Galvani (1737–98) of Italy noticed that detached frog legs twitched when exposed to charge, even though there was no contact between the source of the electricity and the legs. In particular, he noticed that when someone held a metal scalpel to the nerve of a frog's leg, and someone else discharged a Leyden jar nearby, the leg jumped. Galvani was soon convinced that frogs generated what he called "animal electricity" that flowed through their nerves and muscles. He published his results in 1791.

Several years earlier Benjamin Franklin had shown that electricity was associated with thunderstorms, so Galvani decided to see if frog legs would jump in a thunderstorm when a Leyden jar was discharged. Using brass hooks, he attached several legs to an iron railing on a terrace outside his lab. As expected, the legs twitched during the thunderstorm; but to his surprise they continued twitching after the thunderstorm cleared up. Checking further, he found that as long as he had two different metals, such as iron and brass, attached to the legs they would twitch. He also found that several other pairs of metals would, in fact, work just as well.

Above: Batteries such as these are usually 1.5 volts and are used in flashlights and various other electrical devices. Top left: Volta's crown of cups, which he used for creating the first steady current.

VOLTA

The Italian physicist Alessandro Volta heard about Galvani's result in the early 1790s. Within weeks he had repeated most of Galvani's experiments and soon discovered that "animal electricity" was not needed. He showed that the only thing needed was two dissimilar metals and a moist conductor (which replaced the frog leg). He also decided that if a single bimetallic strip and a moist conductor worked, two or more in a series would work even better. So he set up an apparatus consisting of an array of bimetallic strips and moist conductors. His first moist conductor was a cup of salty water. He dipped one end of the bimetallic strip, which was made up of silver and tin, into it, and the other end into a second cup, and so on for the entire array. If a conducting wire was brought from the tin of the last cup to the silver in the first, a spark was produced. Furthermore, if the conducting wire was connected from the tin of the last cup to the silver of the first, a continuous current flowed through the wire.

The apparatus, however, was very cumbersome, and Volta soon improved it. He decided to build what is now called a pile. It was made up of disks of silver and zinc about an inch in diameter piled one on top of the other, with cardboard disks that had been soaked in salt water between them. He continued the sequence until he had a pile of about twenty of these groups of disks, and when he connected a wire from the top disk (silver) to the bottom one (zinc), a continuous current flowed. This was the first "battery" and was the basis of the batteries we use today.

With this discovery it became possible to study continuous flow of electric current for the first time. Volta published his results in 1800. The two terminals of the battery became known as electrodes, with the positive electrode being called the anode and the negative electrode the cathode.

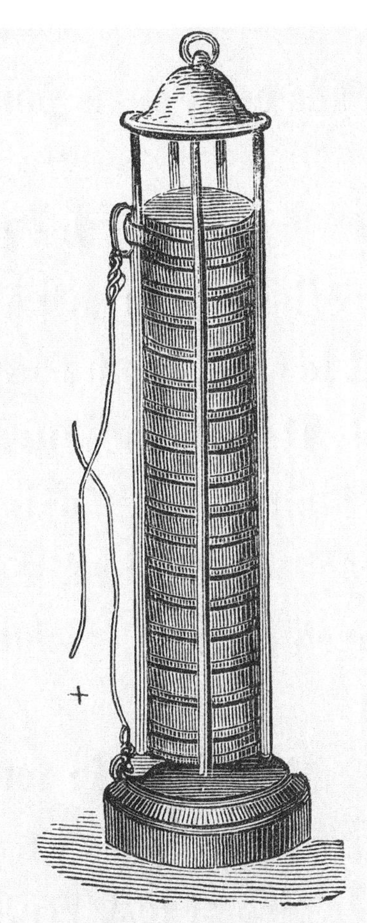

The first voltaic battery made of alternating metal layers and saline-soaked leather pads.

ELECTROLYSIS

Within a few weeks of Volta's discovery two English scientists, William Nicholson (1753–1815) and Anthony Carlisle (1768–1842), passed an electric current through water and found that the water broke up into hydrogen and oxygen. The process was called electrolysis. Much of the early work on electrolysis was done by Michael Faraday. He found that if the two electrodes from a battery were both immersed in a fluid, electricity would flow sometimes, but not always. If the electrodes were immersed in sulfuric acid, for example, current would flow, but in a dilute solution of sugar water it would not. Faraday named liquid conductors, such as sulfuric acid, *electrolytes*; nonconducting liquids were nonelectrolytes. He also noticed that when an electric current was passed through an electrolyte, various elements appeared at the poles. In particular, if these elements were metals they would coat the electrode. We now refer to this as electroplating.

Circuits

With the flow of charge, a new unit was needed, and since charge was measured in coulombs, the rate of flow of charge, or current, would be measured in coulombs per second. One coulomb per second was defined as one ampere, named for the French physicist André-Marie Ampère (1775–1836). It was also soon noticed that the current that flowed along a wire between two points depended on the resistance in the wire. Resistance is measured in terms of ohms, named for the German physicist Georg Ohm (1789–1854). Using an analogy of current in water, resistance can be seen as a measure of resistance to flow. If water in a pipe, for example, encounters a restriction (a sudden narrowing of the pipe, for instance), the flow is reduced. In the same way, if electrical current encounters a resistance, the flow is reduced.

Above: Georg Simon Ohm, the German physicist who discovered the relationship between current, voltage, and resistance, now known as Ohm's law. Top left: An electric meter.

Earlier we saw that the flow rate between two points also depends on the potential difference between the two points. This is, in effect, the pressure that is pushing the current. It is measured in volts. We therefore have three major components of electricity: voltage, current, and resistance. The relationship between them is called Ohm's law, and it is given as

$$V = IR$$

where V is voltage, I is current, and R is resistance. It is easy to see from this that we can determine the current in a circuit as $I = V/R$, where a circuit is a loop containing a battery and a resistance (connected by a conducting wire).

ELECTRICAL POWER

It takes energy to keep the current going against the resistance in a circuit. We saw previously that the work per unit charge in transporting electricity from one point to another is defined as the potential difference. Therefore we can say that work = V, and we know that the rate of doing work is power. Since work has the same units as energy, namely joules (in the mks system), we can define the unit of power as a joule/sec.

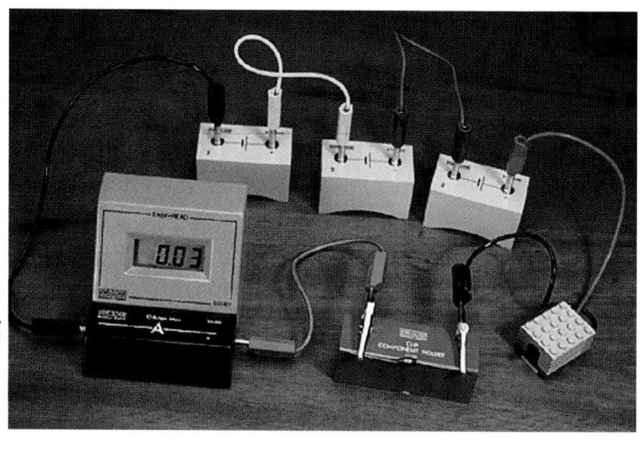

A simple electric circuit in a laboratory. The circuit consists of three batteries that are powering an electric motor. A resistor and ammeter are also in the circuit.

But 1 joule/sec is defined as a watt. Note also that since a joule is a coulomb-volt, that 1 watt is also 1 volt-coulomb/sec or 1 volt-amp. This tells us that we can write the formula for power as

$$P = IV$$

where P is power, I is current, and V is voltage. Furthermore, using Ohm's law, this becomes

$$P = I^2R$$

As an example, consider a 60-watt lamp. We know that the voltage driving it is 120 volts. The current passing through the filament is therefore $I = 60/120 = \frac{1}{2}$ amp, and the resistance is $R = V/I = 120/\frac{1}{2} = 240$ ohms.

In relation to household electricity, the kilowatt is a much more common unit than the watt; it is 1,000 watts. Also, in terms of energy we have the kilowatt-second, but again a more common unit in relation to electric bills is the kilowatt-hour.

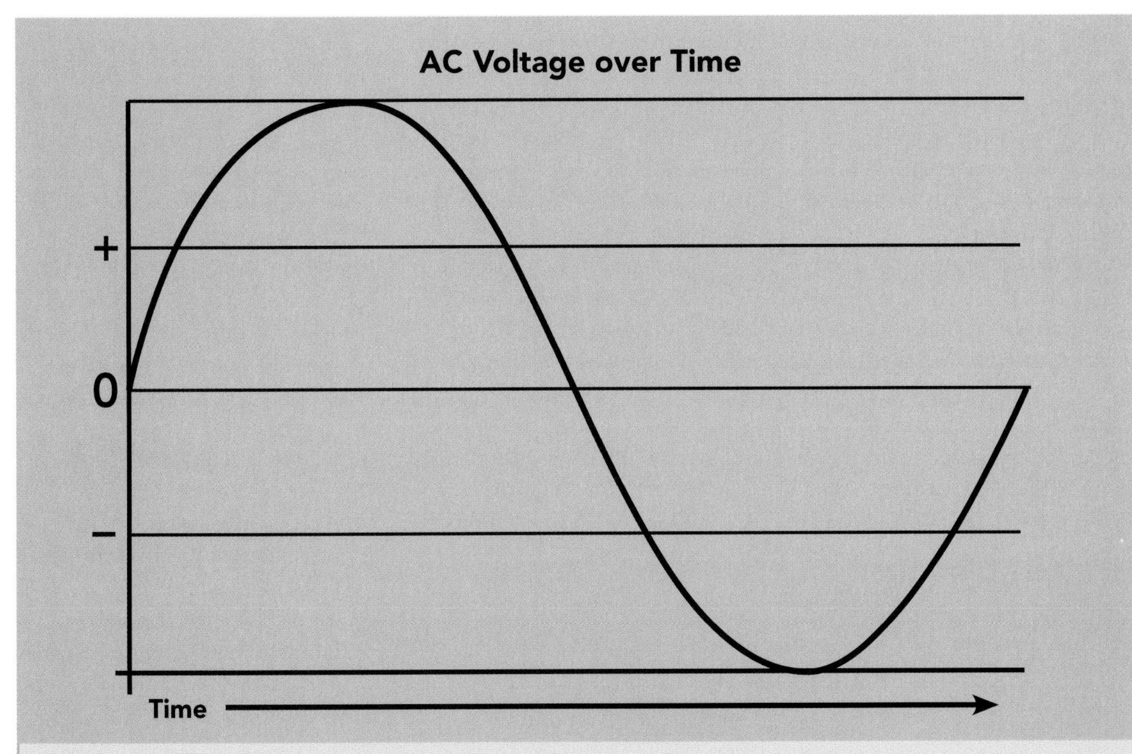

An alternating current (AC) power source changes constantly in amplitude and regularly changes polarity, which can be plotted as a sine wave.

DIRECT AND ALTERNATING CURRENTS

So far we have discussed only steady, or direct, current (DC). But in practice, another type of current is much more common; it is referred to as alternating current (AC). In this case the current flows in one direction for a fraction of a second, then flows in the opposite direction for the same amount of time. It does this over and over. If it is 60-cycle AC, for example, it will alternate like this 60 times each second. If we plot current as a function of time, it would look like a "sine curve" (see figure). At first glance, it might appear that this would not be useful compared with DC current. But it has turned out that AC is used much more commonly than DC. Household current, for example, is AC.

One of the advantages of AC, as we will see a little later, is that AC generators are simpler in design than DC generators. One of the initial problems with AC was that it was much more complicated mathematically—in other words, AC theory was more complicated than DC theory. But this was soon overcome.

Series and Parallel Circuits

Two types of circuits are used in electricity: series and parallel. The simplest is the series circuit. In any series circuit there are three main components: a battery or source of voltage, a load or resistance, and conducting wire that connects them (see figure opposite).

So far we have considered the current to be a fluid of some sort; but as we will see in the next chapter, what actually flows in the wire is negatively charged particles called electrons. These electrons are attracted to a positive charge, and therefore they move toward the positive pole of the battery. But as we saw earlier, Benjamin Franklin suggested that current flows from positive to negative, which is opposite the flow of the electrons, and this is the direction that is still assumed for the flow of current (it is called the conventional current direction).

In the circuit shown we have a battery, which we can assume has a voltage of 12 volts, and a resistance, which can take many forms (a lightbulb is a good example). Assume it has a resistance of 6 ohms; the current that flows will be, according to Ohm's law, $I = 12/6 = 2$ amps.

Now assume we have several resistances in our circuit; call them R_1, R_2, and R_3. Assume they are all linked in series. How do we determine the current in this case? As it turns out, resistances in series add, so if the three resistances were 2, 4, and 6 ohms, we would have a total of 12 ohms and could again determine the current from Ohm's law. It would be 1 amp.

It is also important to note that the total voltage drop across the three resistances has to equal the voltage of the battery. This was not a problem with one resistance, but with three resistances it is important to know the voltage drop across each resistance. Again, this can be calculated from Ohm's law. Across the 2-ohm resistance we get $V = IR = 2$ volts. Similarly, across the 4-ohm resistance it is 4 volts, and across the 6-ohm resistance it is 6 volts. Note that the three add up to 12 volts, the voltage of the battery.

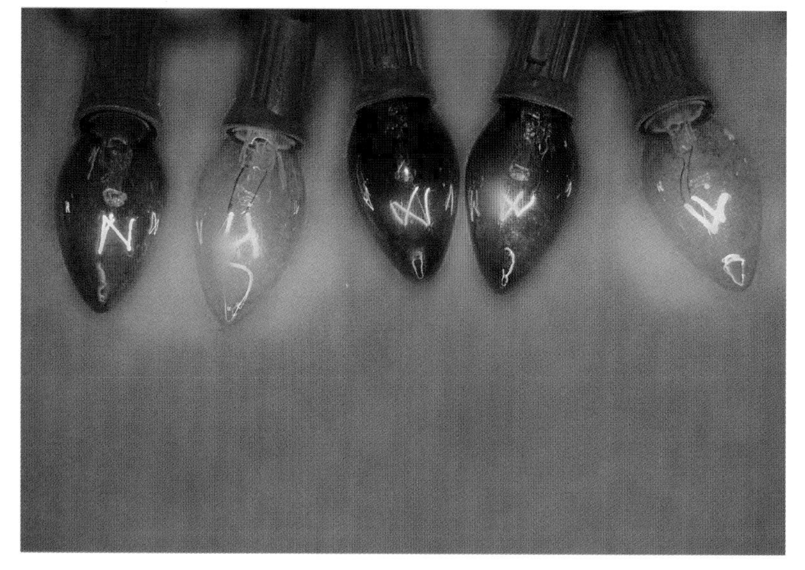

Above: Older Christmas lights were frequently wired in series. If one light burnt out, the whole string went out. Top left: An incandescent light. The filament acts as a resistor.

PARALLEL CIRCUITS

The second type of circuit is the parallel circuit. An example is shown in the figure. The current in the main line now splits into two separate currents. Water flow is again a useful analogy. You know that if a water pipe splits into

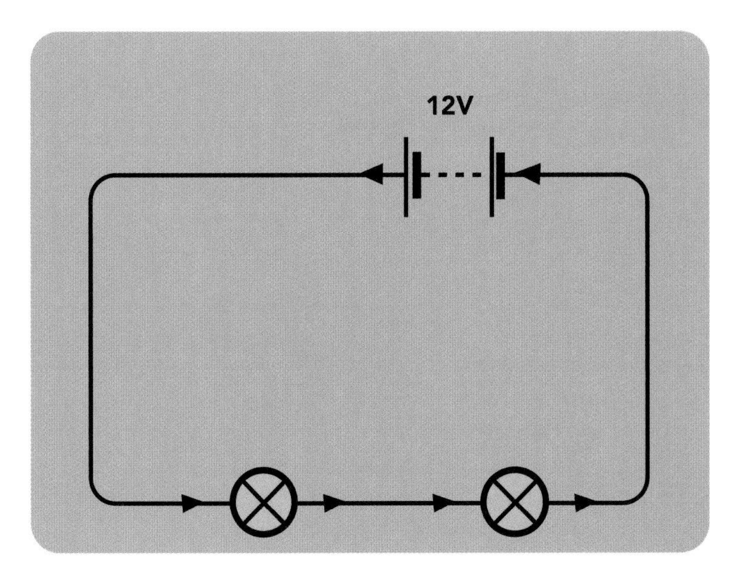

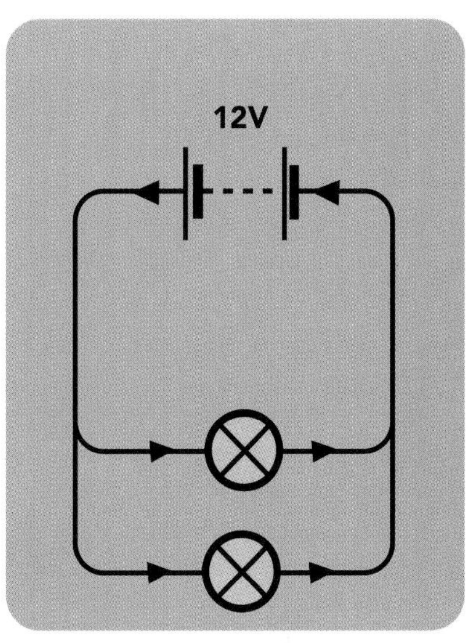

These two diagrams show a series circuit showing battery and resistors (left) and a parallel circuit showing batteries and resistors (right). Arrows indicate direction of current.

two pipes, part of the water goes into one of the pipes and the rest goes into the other. We can still use Ohm's law to calculate the currents in the two branches, but first we have to know how to "add" the resistances.

The rule for resistances in parallel is

$$R = 1/(1/R_1 + 1/R_2)$$

Therefore, if R_1 is 4 ohms and R_2 is 6 ohms, we get $R = 1/(1/4 + 1/6) = 2.4$ ohms. The current that flows in the main line is therefore $(12/2.4) = 5$ amps. We can also use Ohm's law to find the current through each of the resistors (the voltage is the same across both of them, namely 12 volts). The current through the 4-ohm resistance is $12/4 = 3$ amps, and the current through the 6-ohm resistance is $12/6 = 2$ amps, which adds to 5 amps, the current in the main line.

A third possibility is a combination of series and parallel in the same circuit. In practice this is not as common as the previous cases.

Electrical current behaves like water. When water in a pipe comes to a junction with two pipes ahead, part of the water goes into one pipe and the other part into the other pipe, depending on the diameters of the pipes.

Electromagnetism

One of the problems that plagued science for many years was the nature of the relationship between electricity and magnetism. There were so many similarities that it seemed the two had to be related, but no one was able to prove it. The Danish physicist Hans Christian Ørsted (1777–1851) became interested in the problem about 1813 and wondered about it for several years. In the early spring of 1820 he was giving a lecture to some of his more advanced students and planned on checking for a possible link using a compass. He had the experiment clear in his mind, but he did not have time to perform it before the lecture.

The compass was, of course, the natural device for checking for a magnetic field, as it responded sensitively to magnetic fields, and if there was such a field associated with a current-carrying wire, the compass would detect it. As the students looked on, Ørsted brought a compass up to a wire that was attached to a battery. At first he kept the compass needle perpendicular to the wire, and nothing happened. Then he moved it parallel to the wire, and suddenly the compass needle swung around so it was perpendicular to the wire. The movement surprised Ørsted, so he stayed after the lecture and repeated the experiment several times until he was convinced that it was valid.

Surprisingly, Ørsted waited three months before he published the results. In the meantime he investigated other aspects of the discovery. He determined that the magnetic field surrounded (circled) the current-carrying wire, and it dropped off with distance. Also, he showed that the field was greater for larger wires and therefore depended on the amount of current that was

Hans Christian Ørsted, the Danish physicist who discovered the relationship between electricity and magnetism.

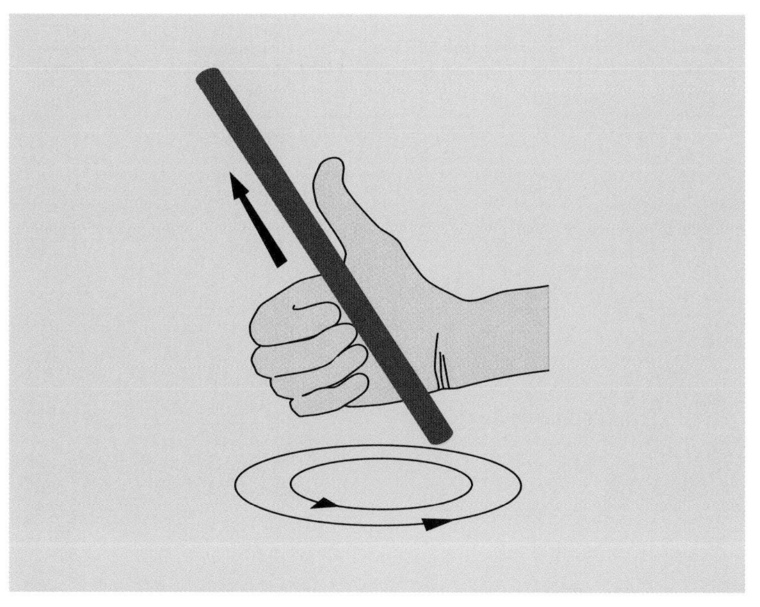

Above: Ampere's right-hand rule. Top left: Electric motor coils. Coils of electrical wire, or solenoids, shown inside a small demonstration electric motor.

flowing. In addition, he performed the experiment under glass, marble, water, and other nonconducting materials and found the field was not affected. Finally, he put a small magnet in the field of the wire and found that it was affected: A current-carrying loop of wire was turned by an external magnet.

Ørsted published his results in July 1820, and they soon caused a sensation throughout Europe and America. There was now proof: Electricity and magnetism were related. An electric current generated a magnetic field. We now refer to the interactions of the two fields as electromagnetism.

SOLENOIDS

The French physicist André-Marie Ampère heard about Ørsted's discovery at a meeting in Paris a few weeks after the discovery. He repeated Ørsted's experiments and went on to experiment with the fields around the wire. He found that if currents were moving in the same direction along two wires, there was a repulsion between them, and if the currents were moving in opposite directions, there was an attraction. He worked out the details of the interactions and showed that they obeyed the inverse square law. He also developed what is now called the "right-hand rule," which says that if you grasp the current-carrying wire with your right hand and close your fingers around it, with your thumb pointing in the direction of the current, your fingers will point in the direction of the magnetic field.

Ampère developed a theory for the magnetism that arises from electricity moving in circular orbits around an axis. He called a wire wound in a spiral that created a magnet a *solenoid*, and he carried out many experiments with wires wound around glass and other materials. From this he developed a mathematical treatment of the interactions between electrical currents and the circular currents around magnets.

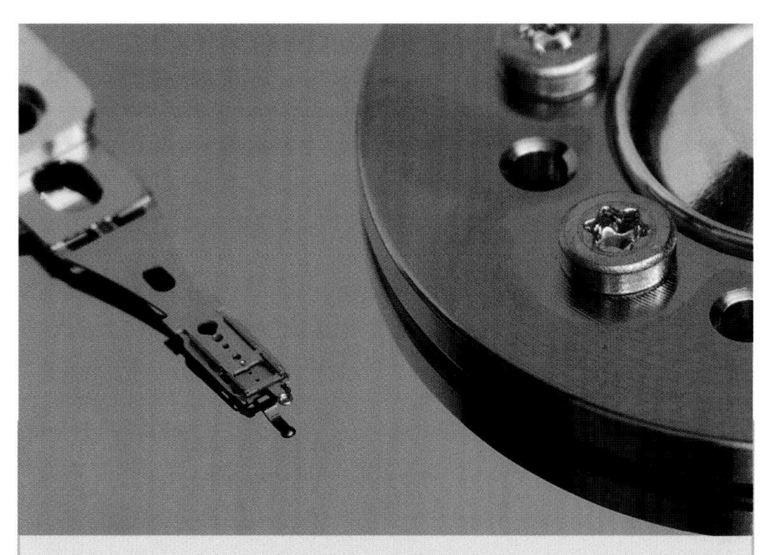

An electromagnetic hard drive reader/writer is an integral part of a computer.

ELECTROMAGNETS

A solenoid creates a magnet similar to a bar magnet, and it was soon discovered that the strength of a solenoid magnet could be increased by placing a bar of iron in it. The English physicist William Sturgeon (1783–1850) was the first to do this; he wrapped copper wire around an iron bar and produced an electromagnet. Furthermore, he noticed that steel could be permanently magnetized, but iron lost its magnetism as soon as the current was shut off. Significant improvements in electromagnets were made by the American physicist Joseph Henry (1797–1878). He used insulated wire for his electromagnet, which allowed many more turns and increased the field. By 1832 he had constructed an electromagnet that could lift more than 3,000 pounds.

Electromagnets are now made with extremely high magnetic fields. To give you some perspective, a small bar magnet usually has a field of a few hundred gauss, while a large one can produce fields of a few thousand gauss. With electromagnets, fields of 50,000 gauss and more are possible. And with magnets made of superconducting material (they have nearly zero resistance), fields of hundreds of thousands of gauss are possible.

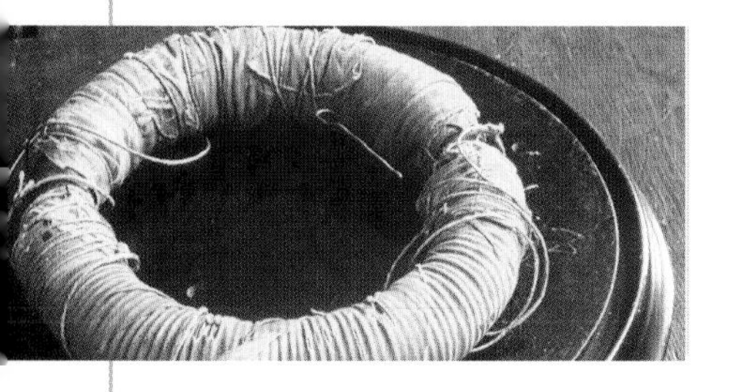

Induction

After Ørsted's discovery that an electric field could produce a magnetic field, it was natural to ask if a magnetic field that was already in existence could produce an electric current. Faraday was particularly interested in this question, and in 1831 he performed an experiment to help find an answer. He began by winding a coil of wire around part of an iron ring; then he attached the ends of the wire to a battery so that he had a solenoid. He also placed a switch in the circuit so he could turn the solenoid off and on.

Above: Michael Faraday, the English physicist who made important discoveries in electricity and magnetism. Top left: The induction coil used by Michael Faraday when he discovered the principle of induction.

On the opposite side of the ring he wound another coil of wire and attached the ends of it to a galvanometer (an instrument that detected current). Faraday thought that switching on the current might generate a current in the second coil. But the result was only a tiny surge in current. The process did not produce a continuous current in the second coil. A tiny current also occurred when he opened the switch. He soon realized that it was not the *existence* of magnetic field lines that created a current, but rather the *motion* of magnetic field lines across a wire that caused it. This is now referred to as electromagnetic induction.

Soon after this experiment, Faraday showed that if he pushed a magnet into a coil of wire, a current was produced in the wire. In particular, the current moved in one direction when the magnet was pushed in, and in the opposite direction when the magnet was pulled out. No current passed if the magnet was stationary.

About the same time, a Russian physicist, Heinrich Lenz (1804–65), added to the discovery. He showed that the induced potential difference in the circuit always opposed the change that produced it. This was called Lenz's law. It means that when a current is created in a coil, its rise causes a potential difference that produces a current in the opposite direction. Similarly, when a current is broken, we get a potential difference that is that is opposite the current; this is referred to as a back voltage. A right-hand rule gives us the direction of the current. Pointing straight ahead with your index finger, raise your thumb and point to the left with your third finger, if your index finger is in the direction of the magnetic field, and the motion of the wire is in the direction of your thumb, the current will be in the direction of your third finger.

THE ELECTRIC GENERATOR

As we saw, an electric current is produced when magnetic lines of force cut across an electric conductor. But the effect is short-lived. How was it possible to create a continuous current? This was accomplished by making what are called "slip rings" that could be attached to the external circuit. Two wires, ending in brushes or sliding contacts, were pressed against the rings. A coil of wire (called an armature) was placed in a magnetic field. When the coil was rotated, an electromagnetic field was generated, causing a current to flow through the slip rings to the external circuit. This was the first electric generator. It converted energy of motion into electrical energy. An external source of energy such as a waterfall or wind was needed to keep the armature rotating.

The same principle can be used to produce a motor. A motor is just an "inverse" generator. Instead of generating electricity through the use of motion, electricity is used to generate motion.

A large power generator.

Transformers convert electricity between high and low voltages.

TRANSFORMERS

Another device of importance that came from Faraday's discovery was the transformer. It is used extensively in modern electrical equipment for changing the voltage of a source. In a car, for example, about 20,000 volts is needed to discharge the spark plugs, but the battery supplies only 12 volts.

In a transformer, wire is wound around a core of soft iron. Several hundred turns of relatively heavy wire are wound on first; this is the primary coil. Around it thousands of turns of fine wire are wound; this is the secondary coil. There is much more resistance in the secondary coil than in the primary coil because there is considerably more wire. When the primary current is turned on and off quickly (usually with an electronic device), the magnetic field builds up and collapses, and the collapsing magnetic field moves across, or cuts, the secondary windings. This induces a high voltage in the secondary circuit. The ratio of the two voltages is the ratio of the number of windings on the primary to the secondary. With a large number in the secondary, we can obviously increase the voltage considerably. Of course, as the voltage goes up, the current will go down. We can also create a step-down transformer in the same way (it goes from high to low voltage).

Maxwell, Light, and Electromagnetism

At the end of the eighteenth century, four major things were known about electricity and magnetism. They were:

1. An electric charge is surrounded by an electric field. The direction of the field is such that like charges repel and unlike charges attract.
2. There is no such thing as an isolated magnetic pole (N and S poles always exist together).
3. A magnetic field can be generated by a changing electric field (or a moving charge).
4. An electric field can be generated by a changing magnetic field.

In the early 1860s James Clerk Maxwell decided to supply a mathematical analysis of the four above observations. In 1864 he published a set of four equations, one for each of the above, that are now known as Maxwell's equations. Since the above principles were already known, and Maxwell merely put them in mathematical form, it might not seem to be a great achievement, but it was. The new equations gave us an unprecedented understanding of electricity and magnetism and allowed us to look at the relationships between the two phenomena in a new light.

Maxwell's work showed was that it was impossible to consider electricity and magnetism in isolation. They were intricately related, and together they formed what is now called an electromagnetic field. Furthermore, oscillation of electric charges produced electromagnetic waves that spread out from the charges. These were waves that had both electric and magnetic fields associated with them. The magnetic field was in one plane, with the electric field perpendicular to it; both were perpendicular to the direction of travel. Maxwell was able to calculate the speed of the electromagnetic waves and found that they traveled at the speed of light. He was sure this was significant, and as a result he suggested that light was electromagnetic waves.

It also seemed to Maxwell that since the charges that created the electromagnetic waves could, in theory, oscillate at any frequency, there should be a large array of electromagnetic waves of different frequencies beyond the shortest and longest wavelengths of light. In other words, there should be wavelengths longer than that of red light and shorter than that of violet light.

In 1873 Maxwell published his famous book, *Treatise on*

Above: A technician at a radio company works with early electric devices in the 1920s. Top left: Simple representation of radio waves being transmitted from an antenna.

Electricity and Magnetism. In it he explained his theory of electromagnetic waves, going into more details on their properties; he also summarized the knowledge of electricity and magnetism to that time.

HERTZ

The German physicist Heinrich Hertz (1857–94) was the first to prove that electromagnetic waves actually existed in nature. He used a relatively simple device for his experiment. It consisted of two loops of wire with gaps in each; small brass knobs were attached at the gaps. According to Maxwell's theory, electromagnetic waves should be emitted as a spark jumped the gap. Hertz therefore set the two coils several feet apart and passed a current through the first coil, making sure it jumped the gap. And indeed a current was produced in the second loop. He was the first to intentionally produce electromagnetic waves. Furthermore, he was able to calculate the wavelength of the radiation, and he showed that it contained both electric and magnetic fields. Hertz also discovered that electrical conductors reflected the waves, and that they could be focused by concave reflectors.

German physicist Heinrich Rudolf Hertz, who discovered radio waves.

Guglielmo Marconi, the Italian physicist who transmitted the first radio message across the Atlantic Ocean.

RADIO WAVES

One form of electromagnetic radiation is radio waves. In 1894 Guglielmo Marconi (1874–1937) of Italy read an article on radio waves and decided to see if he could detect them. He was sure they would be useful in signaling. Using Hertz's technique for producing the waves, he set up a transmitter and receiver. For detecting the waves he used a device called a coherer, which consisted of a container of metal filings. It had been previously discovered that such a container conducted considerable electricity when electromagnetic waves fell on it, and they could be converted to an electrical current. Marconi eventually turned to an antenna for both sending and receiving the waves, and over the years he gradually improved his equipment.

In 1875 he sent and received a signal over a distance of one and a half miles. Within two years he had improved this to 12 miles. At the turn of the century he planned to send a message between England and Newfoundland, but there appeared to be a problem: Electromagnetic waves traveled in straight lines, and the Earth curved. He was pleased to discover, however, that the waves followed the curvature of the Earth. The reason was that the radio waves were reflected, or bounced, from electrically charged particles in the atmosphere.

The Electromagnetic Spectrum

Maxwell's prediction was right. There are electromagnetic waves of both shorter and longer wavelengths than those of visible light. There is a whole spectrum of electromagnetic waves, ranging from very short ones called gamma rays up to very long radio waves. We saw earlier that the range of wavelengths for visible light is from approximately 3,800 Å to 7,600 Å. Beyond these two limits, however, are a large number of other electromagnetic waves of higher and lower wavelengths. Beyond the red end of the spectrum are infrared waves, and beyond the infrared region are microwaves. Similarly, beyond violet light is ultraviolet light, and beyond ultraviolet light are X-rays. Finally, beyond X-rays are gamma rays.

It is important to note that electromagnetic waves carry energy, or more exactly, they *are* a form of energy. In fact, the higher the frequency of the waves, and the smaller the wavelength, the greater the energy. This is why X-rays have much more energy than visible light.

The Electromagnetic Spectrum

10^{-6} nm
10^{-5} nm
10^{-4} nm
10^{-3} nm
10^{-2} nm 1 Å
10^{-1} nm
1 nm
10 nm
100 nm UVIS EUV - 55.8-118 nm UVIS FUV - 110-190 nm
10^3 Nm 1 µm
10 Nm
100 Nm
1000 Nm 1 nm
10 mm 1 cm
10 cm
100 cm 1 m
10 m
100 m
1000 m 1 km
10 km
100 km
1 Mm
10 Mm
100 Mm

Gamma rays
X-rays
Ultraviolet
Visible light
Near infrared
Far infrared
Microwaves
Radio

Visible Light ~400 nm -~70 nm
Violet
Indigo
Blue
Green
Yellow
Orange
Red

UHF
VHF
HF
MF
LF

Audio

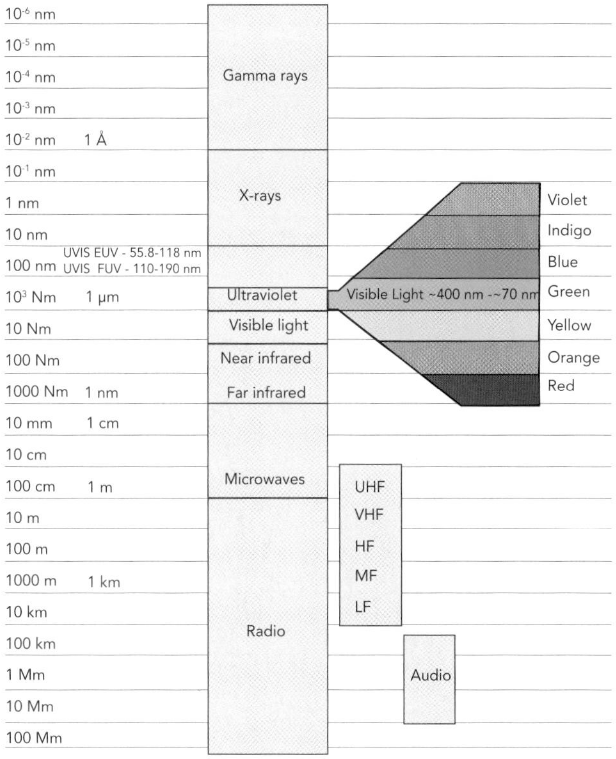

nm=nanometer, Å=angstrom, µm=micrometer, mm=millimeter, cm=centimeter, m=meter, km=kilometer, Mm=Megameter

Above: The electromagnetic spectrum: a plot of electromagnetic waves from very short gamma rays to long radio waves. The wavelengths of the various types of waves are shown to the left. The visible spectrum of colors is shown on the right. All electromagnetic waves travel at the speed of light. Top left: Traffic radar speed cameras are used along many highways.

INFRARED WAVES, MICROWAVES, RADIO WAVES

In 1800, while testing the Sun's spectrum using a thermometer, German-born astronomer William Herschel (1738–1822) found that the temperature rise was the highest beyond red. He concluded that sunlight contained an invisible radiation that was associated with heat. And we now know that the most common source of infrared radiation is heated objects. When you turn on the burner of a stove you feel the heat long before the burner turns red.

The various wavelengths can be referred to either by their wavelengths or their frequencies. The unit of frequency is cycles per second, which we now refer to as hertz, or Hz. A typical frequency of infrared light, for example, is 100,000 MHz (where M is mega, or one million).

Microwaves are familiar to most people because of microwave ovens. The wavelength of these waves extends from about 30 cm to 0.3 cm, or from a frequency of 1,000 to 100,000 MHz. Microwaves are generated in a microwave cavity, which is basically a metal box where oscillating electric and magnetic fields are set up. The microwave oven, in fact, is a microwave cavity. The frequency of waves in such an oven is usually about 2,450 MHz. This frequency easily penetrates food and cooks it very quickly; in addition, food heats very uniformly at this frequency. Another use for microwaves is radar. This is because microwaves are reflected well from objects such as cars, airplanes, and so on.

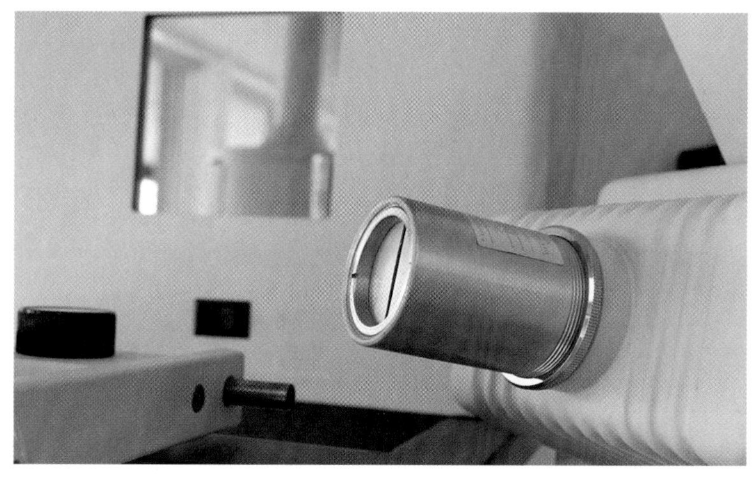

An X-ray machine. Since Röntgen's discovery that X-rays can see through to bony structures, X-ray equipment has become indispensable to many medical practitioners.

The NASA spacecraft Swift *is being used to study gamma-ray bursts in space. It was launched in 2004.*

Radio waves have an exceedingly large range of wavelengths, from a few centimeters to thousands of meters and more. In terms of frequencies, AM radio is usually in the KHz, or kilohertz, and FM radio is in the MHz range.

UV AND X-RAYS

Ultraviolet, or UV, radiation was discovered by the German physicist Johann Ritter (1776–1810) in 1801. The wavelength of UV radiation ranges from 3,600 Å to about 10 Å; this corresponds to frequencies of about 10^{15} Hz.

Beyond UV rays are X-rays. They were discovered by the German physicist Wilhelm Röntgen (1845–1923) in 1895. X-rays have an exceedingly short wavelength and high frequencies.

Milestones in Physics

250 BCE
Archimedes discovers the buoyancy principle that now bears his name.

150 CE
Ptolemy refines the model of the solar system and universe with the Earth at the center. The model eventually becomes known as the Ptolemaic system.

1543
Copernicus introduces a model of the solar system with the Sun at the center. He shows that it explains retrograde motion better than the Ptolemaic system.

1609
Kepler publishes his three laws of planetary motion. In particular, he shows that the planets move in elliptical orbits with the Sun at one focus.

1610
Galileo shows that all objects fall at the same rate if air resistance is neglected. He also makes important contributions to mechanics and builds one of the first telescopes, making several important discoveries with it.

1665
Newton introduces his three laws of motion and his law of gravity.

1678
Christian Huygens proposes the wave theory of light.

1752
Benjamin Franklin flies a kite in a storm and shows that lightning is electrical in nature.

1791
Luigi Galvani discovers that muscle and nerve cells in frog legs produce electricity.

1800
Alessandro Volta constructs the first storage battery.

1803
John Dalton introduces the idea of atoms.

1803
Thomas Young demonstrates proof of the wave theory of light.

1820
Hans Ørsted discovers that an electric current can produce a magnetic field.

1824
Sadi Carnot gives a formula for the efficiency of a heat engine. His work leads to the science of thermodynamics.

1830–31
Michael Faraday discovers that current can be induced magnetically. He also introduces the idea of magnetic field lines.

1864
James Clerk Maxwell brings together the ideas of electricity and magnetism into four equations that now bear his name. He also predicts the existence of electromagnetic waves.

1887
Albert Michelson and Edward Morley perform an experiment that shows that the speed of light is constant and independent of the motion of the source.

1888
Heinrich Hertz discovers radio waves.

1895
Wilhelm Röntgen discovers X-rays.

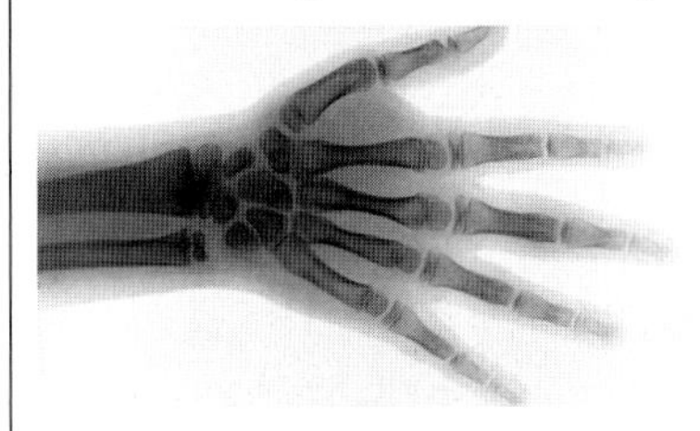

1896
Antoine Henri Becquerel discovers radioactivity.

1897
Sir Joseph John Thomson discovers the electron.

1900
Max Planck derives a formula to explain the emission of radiation from a blackbody. He uses the idea of discreteness, and introduces the first quantum formula.

1903
Antoine Henrie Becquerel and the Curies win the Nobel Prize for their work with radiation phenomena.

1905
Albert Einstein publishes his theory of special relativity, which completely changes our ideas of space and time.

1910
Robert Millikan performs what is now called the Millikan oil drop experiment and determines the charge of the electron. He also determines its mass.

1911
Ernest Rutherford discovers a "nucleus" within the atom. He also later discovers the proton.

1913
Niels Bohr develops the Bohr model of the atom in which the electrons orbit in discrete orbits around the nucleus.

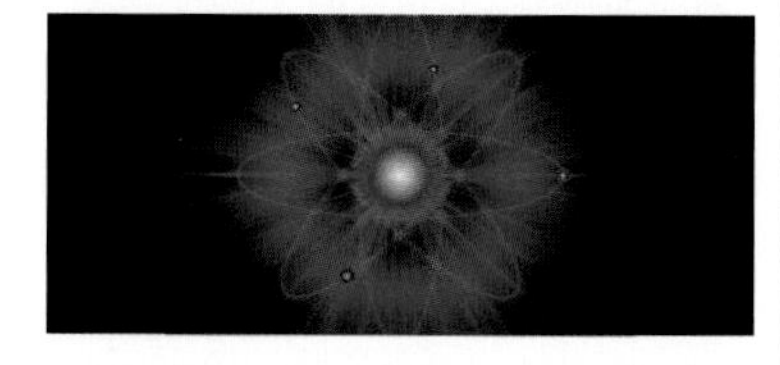

1915
Einstein publishes his general theory of relativity and introduces the idea that gravity is curved space.

1919
Arthur Eddington verifies predictions of general relativity in an eclipse.

1924
Wolfgang Pauli discovers the exclusion principle, which states that no two electrons in the same atom can have the same quantum numbers.

1925
Louis de Broglie discovers the wave nature of particles.

1925
Werner Heisenberg develops the matrix form of quantum mechanics.

1926
Erwin Schrödinger develops the wave form of quantum mechanics.

1927
Werner Heisenberg discovers the uncertainty principle, which states that you cannot simultaneously measure both momentum and position to a high degree of accuracy.

1929
Edwin Hubble discovers that the galaxies are moving away from one another. He introduces the idea of an expanding universe.

1932
James Chadwick discovers the neutron.

1935
Hideki Yukawa predicts the existence of the meson. It is discovered a few years later.

1945
The first atomic bomb is built and exploded. The project is headed by Robert Oppenheimer.

1947
Richard Feynman, Julian Schwinger, and Shin-Ichiro Tomonaga independently formulate quantum electrodynamics (QED). Feynman introduces the diagrams that bear his name.

1953–58
Charles Townes builds the microwave amplification by stimulated emission of radiation device, or MASER, and later designs the light amplification by stimulated emission of radiation device, or LASER.

1962
Murray Gell-Mann introduces the quark theory. George Zweig introduces the similar theory of aces.

1964
Arno Penzias and Robert Wilson discover cosmic background radiation.

1967
Steven Weinberg and Abdus Salam unify the electromagnetism and weak nuclear fields.

1974
Sheldon Glashow creates quantum chromodynamics (QCD). This is the first step toward a Grand Unified Theory (GUT).

1982
John Schwarz and Michael Green develop the idea of strings as fundamental entities in the universe.

1988
Stephen Hawking publishes *A Brief History of Time*, which makes theories on astrophysics, astronomy, and cosmology accessible to the masses.

1995
Ed Witten unifies string theory.

2006
American astronomers John Mather and George Smoot win the Nobel Prize in Physics for their work with the Cosmic Background Explorer (COBE) satellite, which provides support for the big bang theory.

Systems of Units

Measurement is of particular importance in any science. In the study of mechanics, three fundamental quantities are measured—length, mass, and time. These are central to all systems of units. The three systems of units that are in current use are referred to as the mks system, the cgs system, and the fps system.

Fundamental Quantities

The three fundamental, or base, quantities in the mks system are the meter, kilogram, and second. In the cgs system they are the centimeter, gram, and second. Both of these systems are metric systems. The third system is the British or engineering system, usually referred to as the fps system, with the fundamental quantities the foot, pound, and second. The fps system is still used occasionally in engineering, and it is the preferred system for daily life in the United States. Throughout the rest

Old-fashioned analog scales, once used in lab work.

SI BASE UNITS		
	SI BASE UNIT	
Base Quantity	Name	Symbol
length	meter	m
mass	kilogram	kg
time	second	s
electric current	ampere	A
thermodynamic temperature	kelvin	K
amount of substance	mole	mol
luminous intensity	candela	cd

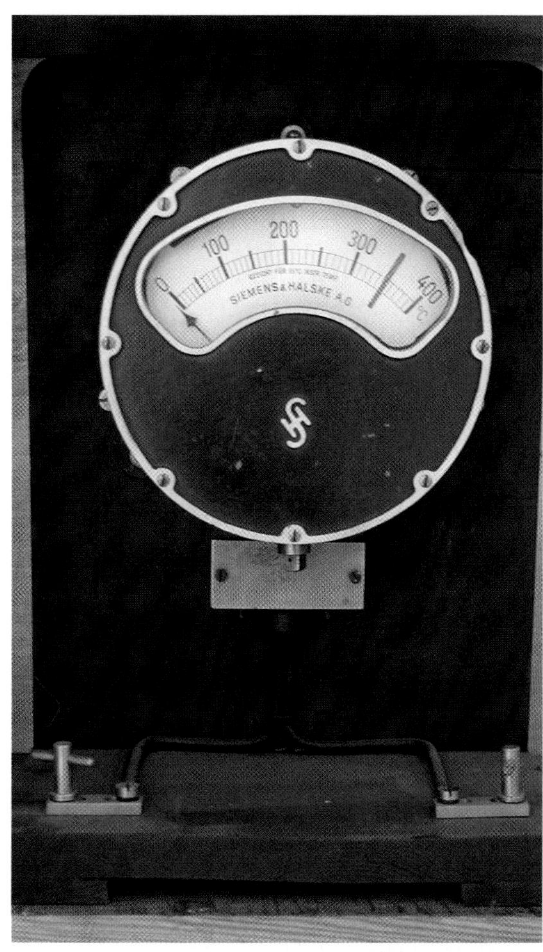

A high-temperature Siemens and Halske thermometer.

of the world, however, the metric system is used.

In 1960 a system called the International System of Units (SI) was introduced throughout the world. It is based on seven fundamental quantities rather than three. Besides the meter, kilogram, and second, it includes the ampere, degrees Kelvin, mole, and the candela. The added four units are, respectively, units of electrical current, temperature, amount of a substance, and luminous intensity. In mks units the unit of force is the newton, and the unit of energy is the joule. The corresponding units in the cgs system are the dyne (force) and the erg (energy).

Derived Quantities

All quantities that are not fundamental are referred to as *derived quantities.* These physical quantities are the result of manipulating the fundamental quantities through multiplication and division—using two or more fundamental quantities, as in meters/second, or the repetition of one fundamental quantity, as in seconds squared.

To illustrate, derived quantities such as velocity and acceleration can be reduced to combinations of length and time. The units of velocity (v) where L is the unit of length and T is the unit of time are:

v = L/T = meters/seconds = m/sec

The units of acceleration (a) would be expressed as follows:

a = L/T^2 = meters/seconds2 = m/sec2.

EXAMPLES OF SI DERIVED UNITS

	SI DERIVED UNIT	
Derived Quantity	Name	Symbol
area	square meter	m^2
volume	cubic meter	m^3
speed, velocity	meter per second	m/s
acceleration	meter per second squared	m/s^2
mass density	kilogram per cubic meter	kg/m^3
specific volumes	cubic meter per kilogram	m^3/kg
current density	ampere per square meter	A/m^2
magnetic field strength	ampere per meter	A/m
amount-of-substance concentration	mole per cubic meter	mol/m^3
luminance	candela per square meter	cd/m^2

How Physics Affects You

Physics is everywhere. It describes the world around us and makes possible the technologies that we utilize every day. The study of the physical world affects you in more ways than you may think.

Traffic in New York City.

When you get up in the morning, chances are a clock radio wakes you. This would not be possible without electricity, which was developed by physicists. When you pop something into the microwave for breakfast or check your e-mail you are affected by physics. Physicists discovered microwaves and made many early advances in computer technology.

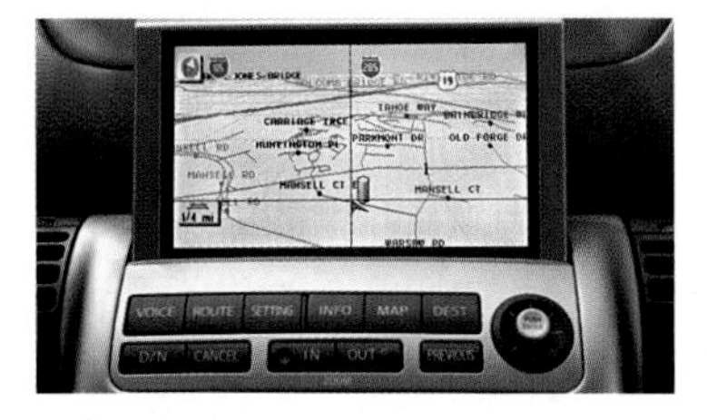

A GPS navigation system.

When you jump in your car and head for work physics is all around you. Speed, acceleration, deceleration, force, torque, and horsepower are all important concepts in physics, and they all apply to cars. The combustion engine that powers your car was built and increased in efficiency with the help of thermodynamics. The car would not run without an electrical system and battery, and the ignition coil is nothing more than a transformer, a basic device of electricity and magnetism. It was physicists' understanding of friction that makes the braking system work. The compact disc player, cell phone, and global position system (GPS) in the car all came about as a result of physics.

When you go shopping, the clerk scans barcodes with a laser, which was invented by a physicist. Ophthalmologists use lasers to repair detached retinas,

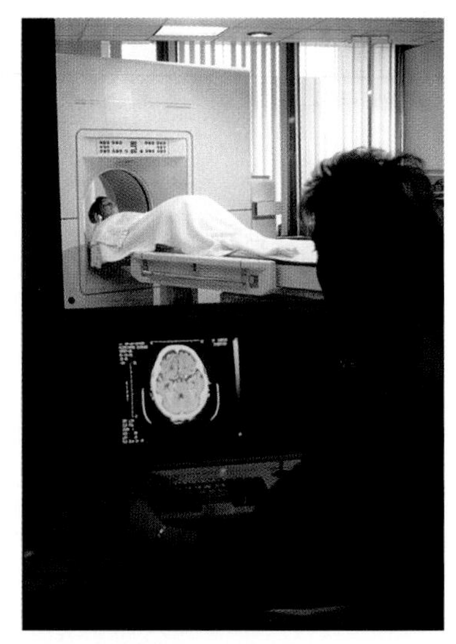

Above: A doctor monitoring an MRI test. Below: A radio communication tower.

destroy tumors in the eye, and for cataract surgery. Surgeons use them to cut out tumors, and in various types of heart and artery operations. Dermatologists use lasers for removing various types of skin lesions, and in the treatment of skin cancer. Lasers are also used extensively in the communications field. They have become particularly useful to the development of fiber optics.

Other advancements in physics have benefited modern medicine. X-rays and the model of the DNA molecule were developed by physicists. Nuclear magnetic resonance, or NMR, which is an important tool of physics, is used extensively for locating tumors in the body, studying blood flow, and examining the brain.

Nuclear medicine involves the use of radioisotopes, radiochemicals, and other technologies for early detection of cancer, the treatment of cancer, and brain imagery. The benefits of nuclear physics, however, are not limited to medicine. Nuclear reactors now supply a large portion of the energy used in the world. We also have nuclear submarines, nuclear fusion reactors, and nuclear batteries.

Satellites play an important role in our civilization, and it was through our understanding of gravity and orbital dynamics that we were able to put satellites in orbit. Satellites are now central to communication: cell phones, television, and GPS all depend on satellites.

The World Wide Web was developed by physicists to speed communications around the world between scientists that needed to share data. Some of the most important early developments were made at the European Organization for Nuclear Research (CERN) and the Stanford Linear Accelerator Center.

From the tiniest particles of matter to the workings of the universe, physics explains how things work and pushes technology into the future.

Above: Using wireless radio technology, computers can access the Internet from remote locations. Below: A surgeon using a laser to perform noninvasive surgery. Bottom: Physicists developed the double-helix model of DNA.

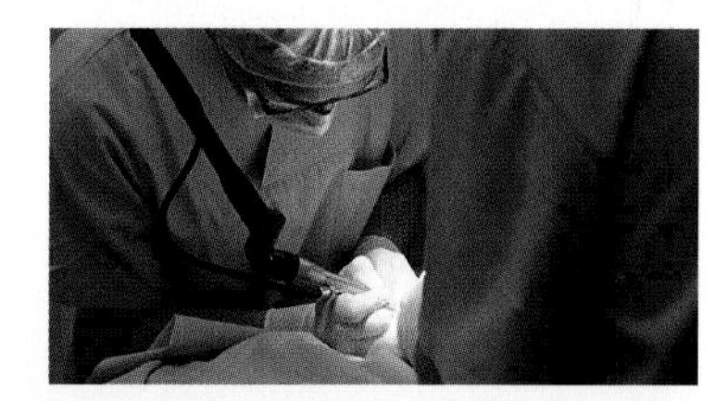

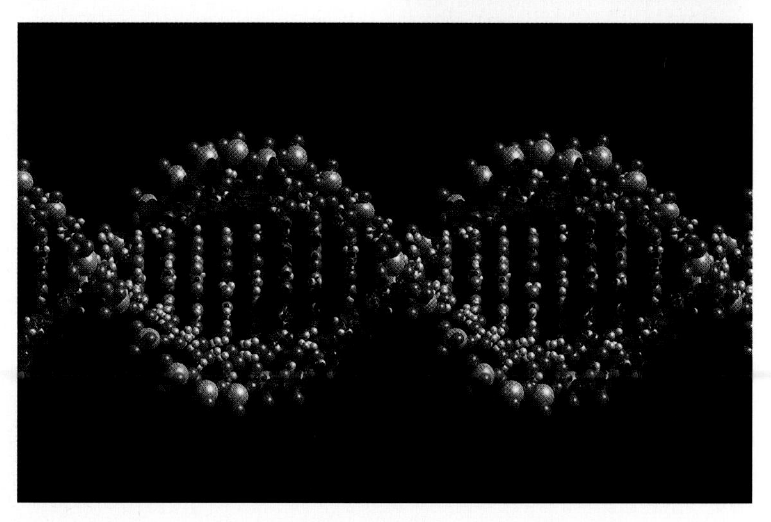

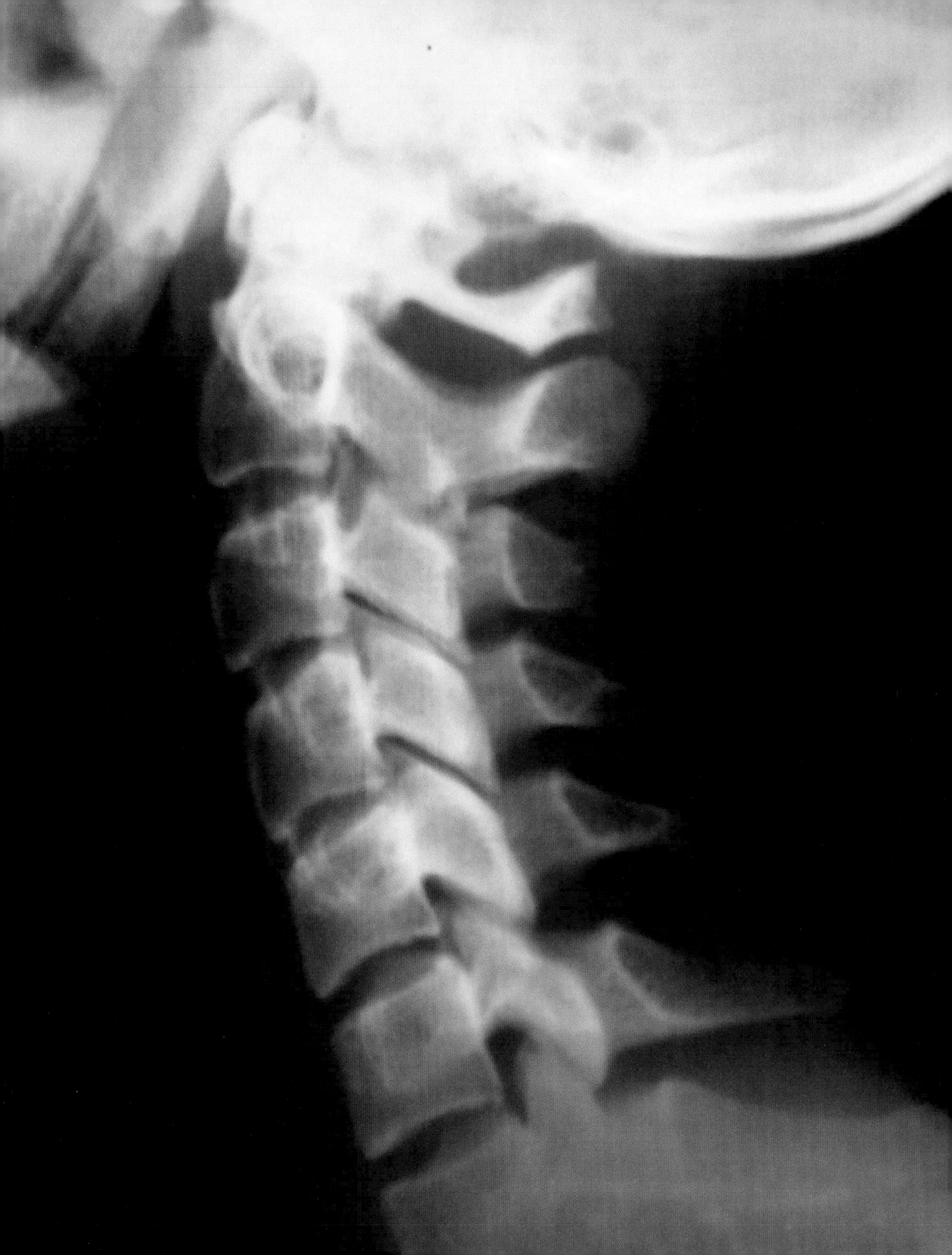

CHAPTER 8

RADIOACTIVITY AND EARLY ATOMIC THEORY

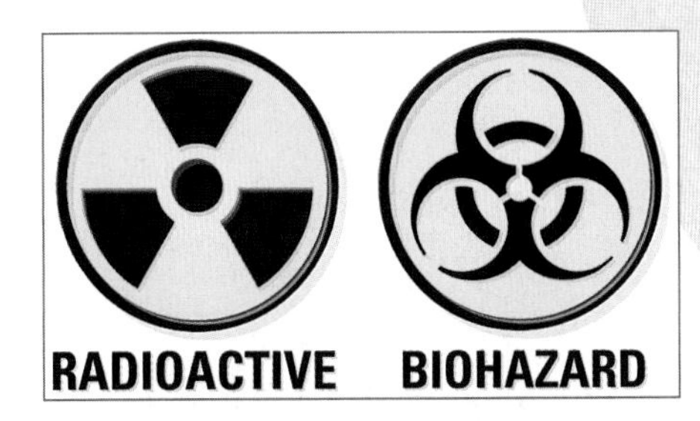

Left: Wilhelm Röntgen's work with cathode rays led to his discovery of the X-ray. Top: When radioactive materials were first discovered, the individuals handling them were not aware of the danger involved. Bottom: Cathode rays were of great interest to physicists working in the early nineteenth century. Cathode ray tubes are the main components in television and computer monitors.

By the late 1800s a feeling of pessimism and gloom had set in within the physics community. Several years passed with no significant breakthroughs, and many scientists were starting to believe that there was little future for anyone interested in physics. The glory days of Newton, Faraday, Maxwell, and others were over. It is perhaps ironic that physics was about to enter what can truly be described as a golden age. Within a few years, exciting discoveries reenergized the scientific community, marking the start of a long chain of events that would completely change physics.

The new era started in 1895 with the discovery of a strange type of radiation. The discovery was made at the University of Würzburg in Bavaria by the head of the department of physics, Wilhelm Conrad Röntgen. Like many physicists of the time, he was working on cathode rays (they are rays that pass through an evacuated tube when a large potential difference is applied to two electrodes in the tube). Röntgen's research would lead him to find the penetrating rays he called X-rays.

Radiation

Röntgen was particularly interested in a glow, called luminescence, that occurred in certain chemicals, and he wanted to see if cathode rays would cause luminescence in the chemicals he was working on. The glow was so dim, however, that it had to be observed in a darkened room.

Röntgen was looking for the effect in late 1895 when he noticed that a glow came from a sheet of paper across the room that was coated with platinocyanide. This was strange, because the light from the cathode rays was blocked off, yet when he turned the tube off, the glow disappeared. Some other type of radiation was being emitted by the cathode ray tube, and although it was invisible, it was highly penetrating.

Looking carefully at the cathode ray tube, Röntgen found that the radiation was coming from the point where the cathode rays were hitting the glass. He studied the new radiation in detail over the next few weeks and determined that it could ionize gases (create charged particles in them), and it did not respond to either electric or magnetic fields. Furthermore, it was so penetrating that it would pass through pieces of wood and thin sheets of metal. He also found that it passed through the human hand, and that a photograph of the hand using the rays showed its bones. He knew that this radiation would have important medical applications; for example, it would be useful in showing broken bones, or locating bullets in a person's body. Unsure of what to call the new rays, he began referring to them as X-rays. Röntgen announced his discovery in late 1895 and was awarded the Nobel Prize for the discovery in 1901.

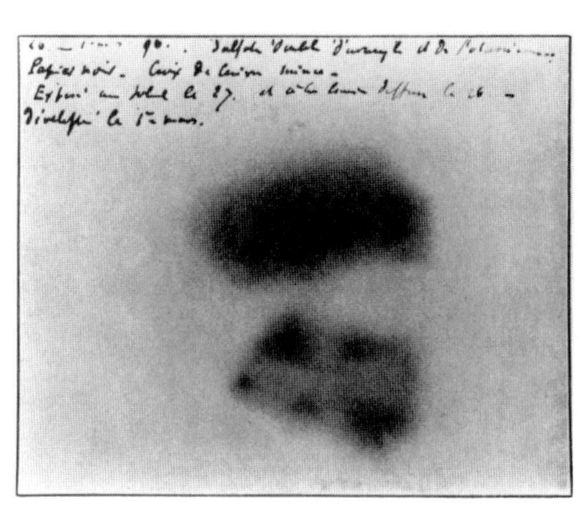

This photograph was generated from Antoine Becquerel's experiment with uranium crystals on photographic paper, which led to his discovery of radioactivity.

Above: Wilhelm Röntgen, the discoverer of X-rays. Top left: pitchblende. Pierre and Marie Curie extracted the radioactive element they called polonium from this ore.

BECQUEREL AND RADIOACTIVITY

Antoine Henri Becquerel (1852–1908) of Paris was studying fluorescence at the time of Röntgen's discovery and wondered if any of the materials he had been working with emitted X-rays. To check, he wrapped photographic paper in heavy black paper and put fluorescent chemicals on top of it, then placed the combination in sunlight, to induce fluorescence in the chemicals. If X-rays were generated, they would penetrate the black paper and fog the photographic paper. The experiment was successful. The photographic film was fogged, but Becquerel knew it was important to verify the result before he announced it. So he got everything ready and waited for another sunny day, but the next few days were all cloudy. Frustrated, he finally decided to develop the film anyway, even though he did not expect anything. To his surprise the film was fogged in the same way it had been when he placed it in the sunlight. It appeared that the Sun had not caused the fogging.

As a final check he took the chemicals and the film in a darkened room, and again the film was fogged in the same way. He concluded that some sort of invisible radiation was coming from the chemicals. It was like X-rays but distinctly different. For example, it was deflected by a magnetic field, he found, and X-rays were not.

THE CURIES

Marie (1867–1934) and Pierre (1859–1906) Curie of Paris soon realized that Becquerel's findings had nothing to do with fluorescence but were a result of the uranium in the chemicals that Becquerel had used. They also found that not only did uranium give off this type of radiation, but thorium did also. Furthermore, they found that an ore called pitchblende was a good source of the radiation, and over the next couple of years they extracted a new element from pitchblende that had the same properties. They called the new element polonium, after the country of Marie's birth, Poland. They also soon began referring to these elements as *radioactive*.

Within a short time they found that pitchblende also radium, that was 900 times as strong radioactively as uranium. In 1902, after several grueling years of sifting through several tons of pitchblende, they were able to isolate one-tenth of a gram of radium. In 1903 they shared the Nobel Prize with Antoine Becquerel for their work.

Pierre and Marie Curie. The couple shared the Nobel Prize with Antoine Becquerel in 1903 for their work with radioactivity.

EXTRACTING RADIUM FROM PITCHBLENDE

Pitchblende is an ore consisting largely of uranium dioxide, but it also contains other substances such as uranium trioxide, lead oxides, thorium, rare earths, and of course, radium. The pitchblende from which Pierre and Marie Curie extracted their radium was obtained from the St. Joachimsthal mine in Bohemia, which was then part of Austria. The Curies were particularly interested in this ore because the uranium had already been extracted from it, and they were not interested in the uranium. The fact that it had already been extracted would save them time.

The story of the extraction of the radium from this ore is now legendary. To get the amount of radium they wanted, they needed several tons of the ore, and each gram of it had to be ground, dissolved, filtered, precipitated, crystallized, and so on to get the radium out. It was a tedious process that took them four years. And perhaps the worst part was that they had no idea how dangerous the material was.

Early Atomism

Two important discoveries had been made, but they were still not well understood. One of the major difficulties was that the ultimate nature of matter was still a mystery. The first prediction of atoms to be taken seriously was that of John Dalton in the early 1800s, and the main reason for this was that he presented considerable evidence for their existence. Still, at this point very little was known about their properties. A discovery that would help science understand these particles better was made in 1869 by the Russian chemist Dmitri Ivanovich Mendeleev (1834–1907). In the course of working with the elements (about 60 were known at the time), he began using cards to list their chemical properties. To his surprise he discovered that if he placed the cards in rows and columns in a certain way, the elements on all the cards in the same columns had similar chemical properties. There were a few problems, but Mendeleev left gaps in the table to account for them. He finally realized that these gaps represented new elements that had not yet been discovered. His insight led to the periodic table of elements that is now used extensively in chemistry and physics.

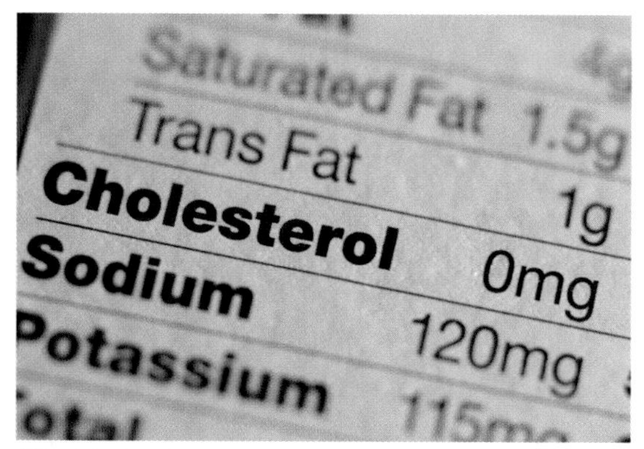

Physicist Michael Faraday identified particles that he called ions. Today, we know that these atoms, or groups of atoms, are necessary for us to live. Sodium ions regulate heart activity.

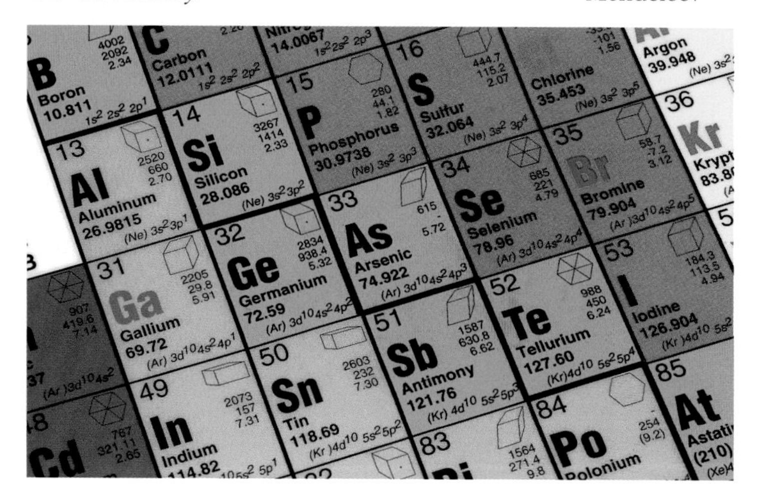

Above: The periodic table of elements was developed by Russian chemist Dmitri Ivanovich Mendeleev. Top left: A simple cup of tea led Albert Einstein to investigate the size of atoms.

FARADAY'S IONS AND CATHODE RAYS

Earlier we saw that Faraday made important contributions to electrolysis. In particular, he noticed that when a current flowed in an electrolyte, certain elements always appeared at the electrodes. This convinced Faraday that "particles" of some type had to be drifting through the liquid. He referred to them as *ions*. He had no idea what they were like, but he showed that some were positively charged and others negatively charged.

A few years later, electrical currents were also observed in cathode ray tubes. As we saw earlier, a cathode ray tube is an evacuated tube with positive and negative terminals implanted in it that allow a voltage source to be attached. The leads are sealed in glass, and a voltage of about 10,000 volts is applied. A fuzzy beam can then be seen traveling down the center of the tube, and if you cut a hole in the anode, the beam passes through it and strikes the glass at the far end of the tube, causing it to glow.

So two phenomena that appeared to show the existence of these particles had been discovered, but in both cases scientists were uncertain just what the particles were. Furthermore, although atoms had been predicted, scientists had no idea how big they were, or how they were made up.

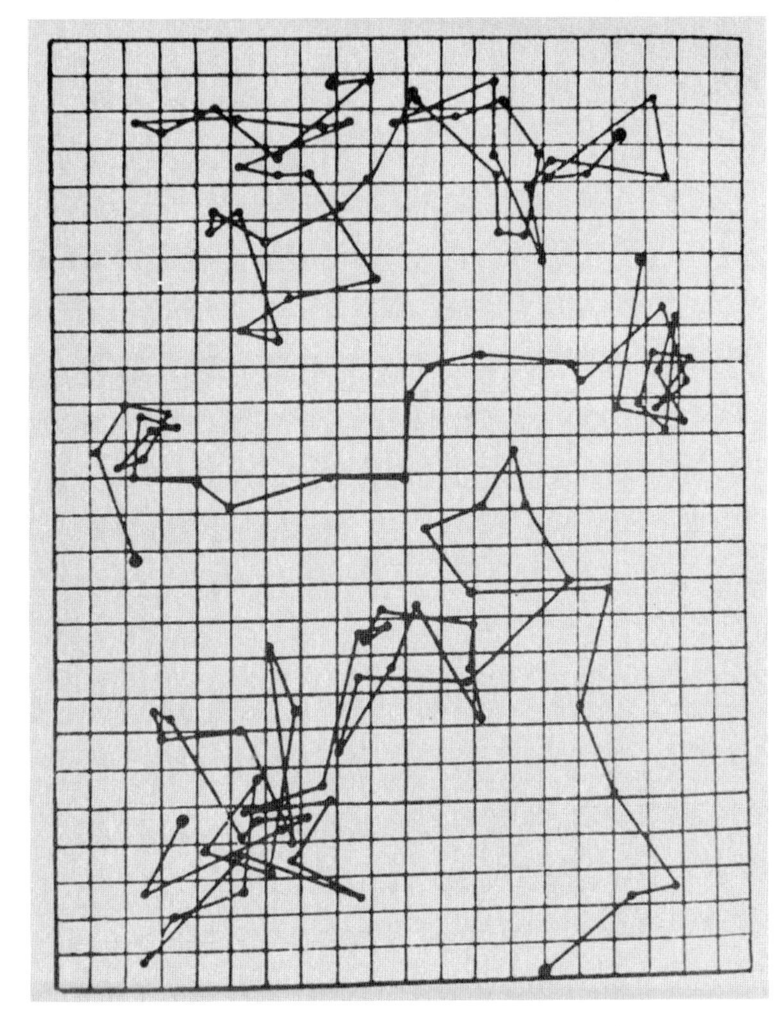

Scottish botanist Robert Brown observed the random motion of pollen grains suspended in water and plotted them on this graph. Called Brownian motion, *this was the first demonstration of the existence of molecules.*

THE SIZE OF THE ATOM

Albert Einstein (1879–1955) was discussing some of the problems associated with the atom with a friend one day at a café in Zurich. Einstein had been looking for a thesis topic and decided a calculation of the size of the atom would make a good project. He was thinking about the topic as he put sugar into his tea; suddenly he realized that the viscosity of the tea changed, and it might depend on the size of the sugar molecule. That evening he looked into the problem in more detail and found that he could calculate the size of the sugar molecule if he knew the viscosity and diffusion coefficient of the tea. The following day he looked them up and had soon determined that the sugar molecule had a diameter of about one ten-thousandth of a centimeter.

Einstein followed this up by showing how the atomic weight (the mass of the atom roughly in terms of the mass of hydrogen) of elements could be determined. About 75 years earlier, a Scottish botanist, Robert Brown (1773–1858), had observed with a microscope a strange trembling motion when he examined tiny grains of pollen on water. The effect was called *Brownian motion*, but no one had explained it. Following up on his previous work, Einstein was able to show that atomic weights could be obtained from a measurement of the motion of the grains. This gave further proof of the existence of atoms.

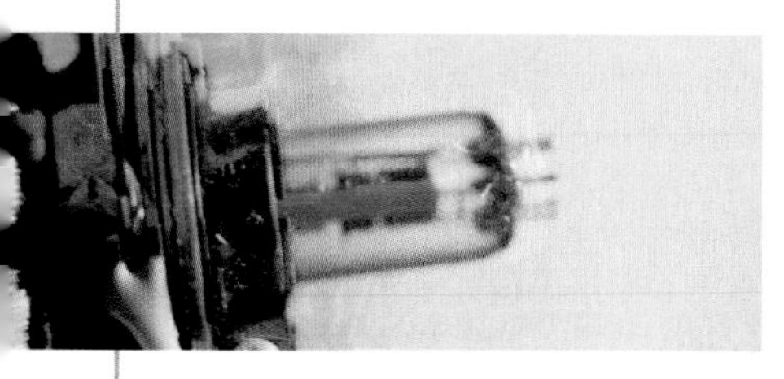

Thomson's Plum Pudding Model

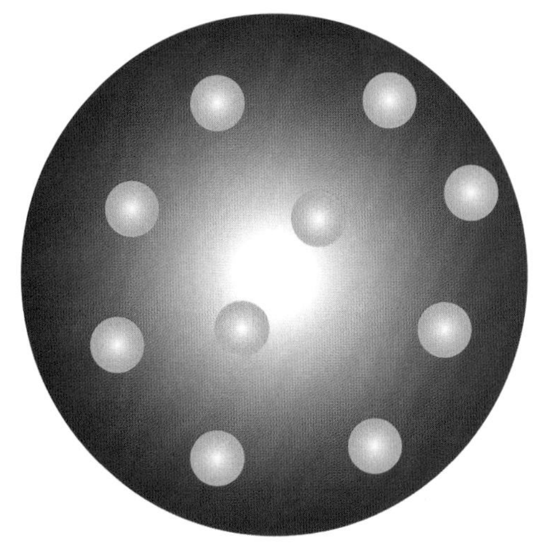

Above: In Thomson's "plum pudding model," each atom is a sphere filled with a positively charged fluid (pink), that he called the pudding. Scattered in this fluid are electrons, known as the plums. Top left: This artwork represents the equipment used when Thomson discovered the electron.

Like many scientists of the time, Sir Joseph John Thomson (1856–1940) of Cambridge University was fascinated by cathode rays. They were known to be negatively charged, and many scientists thought they were particles, but there was no proof. They could easily be deflected by a magnetic field, but they did not appear to be affected by an electric field, and this was a problem. If they were particles, they should have been affected by an electric field. Thomson decided to try much higher electric fields and higher vacuums in his tubes, and as it turned out, this was what was needed. He was finally able to deflect the rays.

By comparing the deflections in electric and magnetic fields he was able to calculate the velocity of the particles, and to his surprise it was about one-tenth the speed of light. No other particles had ever been seen with velocities this high. Once their speed was known, Thomson was able to use the amount of deflection to determine their acceleration, and from it he obtained the ratio of their charge to mass (e/m, where e is charge, and m is mass). It was 770 times higher than the same ratio for a charged hydrogen ion. This meant that if the charge was the same on the two particles (and it was soon proved that it was), the particle of the cathode ray tube was 770 times lighter than hydrogen. (Thomson's value was only approximate; it was later shown to be 1,836 times lighter.)

Sir Joseph John Thomson, who is credited with proving the existence of electrons.

In 1834 the Irish physicist George Stoney (1826–1911) had suggested the name *electron* for the smallest negatively charged particle, and although Thomson objected to it at first, it was eventually used.

THE PLUM PUDDING MODEL

Thomson soon realized that the electron was not only the particle of electric current but was also part of the atom, and within a short time he put forward a model of the atom that he referred to as the "plum pudding model." In it the electrons were embedded in a cloud of positive charge, held in place by

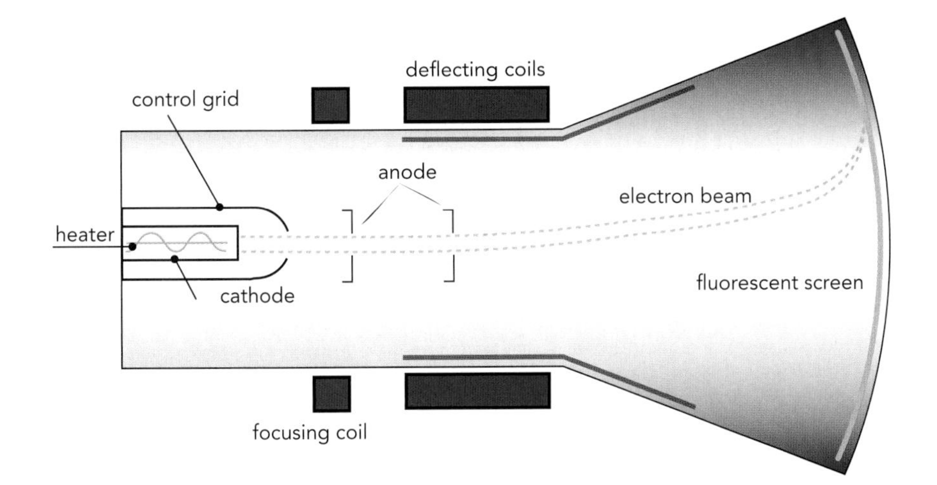

A cathode ray tube is a vacuum tube with an electron gun at one end and a fluorescent screen at the other.

a balance between their negative charge and the positive charge of the cloud. One of the things his model had to explain was the emission of light predicted by Maxwell's theory, and he was able to give a reasonable explanation of it. He pointed out that an outside disturbance, such as a collision, would cause the electrons to vibrate around their equilibrium positions, and as they vibrated they would give off light of the correct frequency.

But Thomson's model was not accepted by everyone. A number of scientists preferred a "planetary" model where the electrons were in orbit. The size of the atom was known to be approximately 10^{-8} cm, and calculations showed that orbits in this range would give the correct frequency to produce visible light. Unfortunately, there was a serious problem: There was no way to stop the emission. The atom would radiate continuously. Furthermore, as the electrons lost radiant energy they would gradually spiral in to the center of the atom. In fact, they would do this in about a millionth of a second, and this was obviously a serious problem.

Robert Millikan's experiment with oil droplets determined the charge and mass of electrons.

MILLIKAN OIL DROP EXPERIMENT

The ratio *e/m* for the electron had been determined by Thomson, but *e* and *m* were not known separately. The American physicist Robert Millikan (1868–1953) of the University of Chicago set out to determine them in 1911. He used an ingenious setup consisting of two horizontal plates, separated by about 1.6 cm; the upper plate was charged positively and had tiny holes drilled in it. To create tiny droplets, Millikan sprayed a fine mist of oil in the region above the plates. Occasionally a small amount of oil would make its way through one of the holes and appear as a droplet in the space between the plates, where it could be seen in an eyepiece. By measuring the rate of fall of the droplet, Millikan was able to determine its mass. Millikan improved the experiment by exposing the mist to X-rays, which ionized it, and when an oil droplet was attached to an ion he had better control of it. In fact, by adjusting the field between the plates, he could move the droplet up and down at will. This allowed him to calculate the charge on it very accurately. Some droplets had more than one ion on them, so a plot had to be made to get the result for a single charge. In this way he found the charge of the electron to be 1.602×10^{-19} coulombs, and its mass to be 9.11×10^{-31} kg.

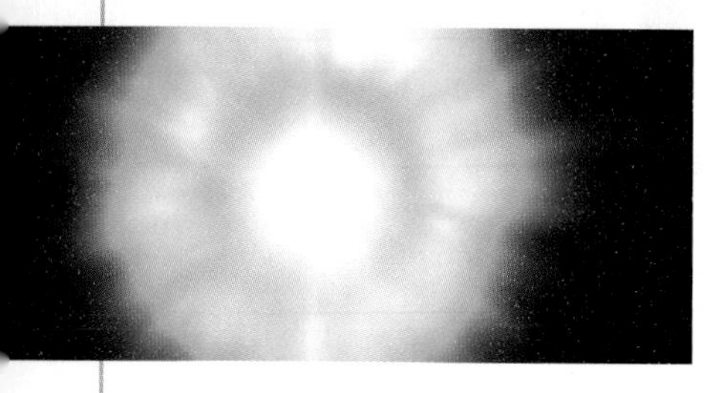

Rutherford and the Nucleus

In 1895 Ernest Rutherford (1871–1937) came from New Zealand to work in J. J. Thomson's laboratory at Cambridge University. He had been in the lab for only a year when Röntgen discovered X-rays. Like most others, Rutherford was amazed by the discovery and soon began investigating the properties of the strange new rays using electric and magnetic fields. Then came Becquerel's discovery of radioactivity, and Rutherford immediately turned to it. He soon determined that two types of radiation were being given off by uranium. He called them alpha and beta rays. Alpha rays were later shown to be ionized helium atoms, and beta rays were shown to be high-speed electrons, so neither was actually radiation. Later it was shown that gamma rays are also emitted, which are radiation.

THE RUTHERFORD-MARSDEN EXPERIMENT

Rutherford's assistant, Hans Geiger (1882–1945), came to him one day looking for a project for a 20-year-old undergraduate student named Ernest Marsden. Rutherford suggested an experiment in which alpha particles were projected at thin sheets of gold foil. The idea was to see how the particles would be deflected when they struck the gold foil. Neither Geiger nor Rutherford expected much from the experiment, as the alpha particles were far heavier than anything in the gold foil. They were sure it would be like firing cannonballs into a swarm of bees. As expected, most of the

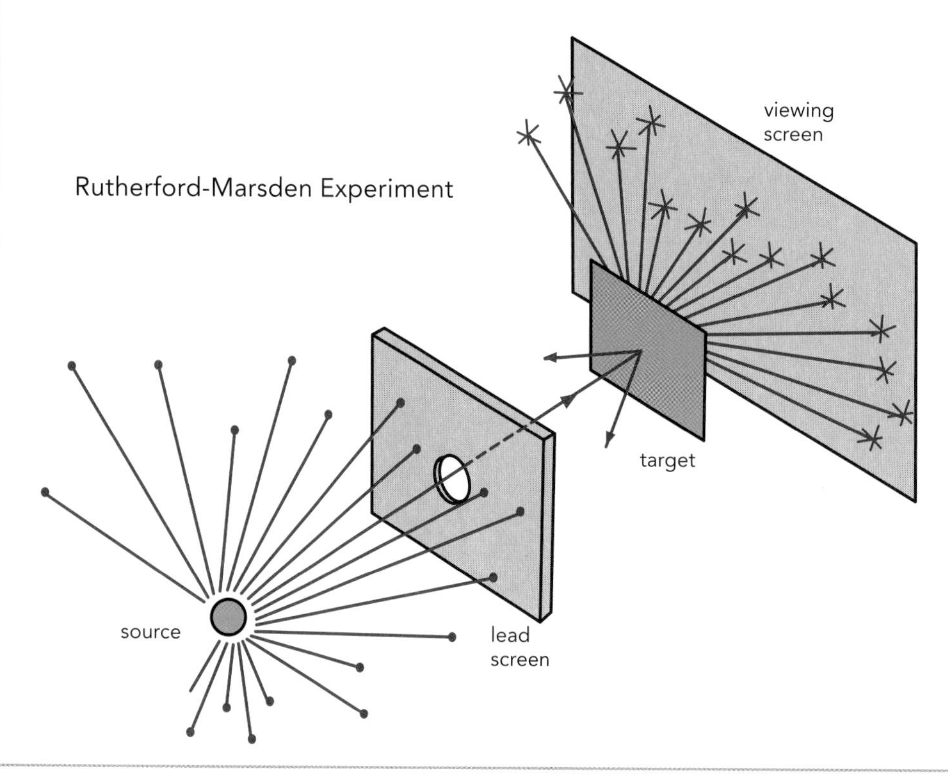

Left: In this experiment, different metal foils were bombarded with alpha particles. The resulting scattered particles were counted, using a microscope. Rutherford observed that a small portion of the particles were deflected, which indicated a small, concentrated positive charge. Top left: European Space Agency rendering of a gamma-ray burst.

alpha particles went through the foil with little deflection, but to their surprise a few underwent large deflections.

Rutherford and Geiger were intrigued by the strange result. Rutherford pondered the experiment for almost two years before he realized what was going on. He finally determined that the large deflections had to be due to an encounter with a tiny "nucleus" at the center of the atom, and this nucleus had to contain virtually all the mass of the atom, even though it had a tiny diameter—only one ten-thousandth that of the overall atom. The immediate consequence of this was that Thomson's plum pudding model was ruled out; but Rutherford did not speculate on an alternative. As far as he was concerned, the only certain thing was that the atom had a massive but very tiny nucleus with a positive charge. He left further speculation to the theorists.

Sir Ernest Rutherford (right), photographed at the Cavendish laboratory at Cambridge University.

Artist's conception of the Rutherford model of the atom.

PROTONS AND NEUTRONS

Rutherford's nucleus was not the only thing with a positive charge. The positive hydrogen ion, for example, was found to have a particularly high *e/m* ratio, and if its charge was equal to that of the electron (but opposite in sign) its mass had to be 1,836 times as great as the electron's. In 1920 Rutherford suggested that the hydrogen nucleus be given the name *proton*. Atoms were therefore made up of two types of particles: electrons and protons.

But by 1930 evidence was mounting that there was a third particle in the atom. When beryllium, for example, was bombarded by alpha particles, a particle was emitted that was not affected by a magnetic field. At first it was thought that it might be a form of radiation, but it did not ionize as other energetic radiations did. Furthermore, when a substance such as paraffin was put in its path, protons were knocked out of the paraffin. The phenomenon was finally explained in 1932 by the English physicist James Chadwick (1891–1974). He predicted the existence of an electrically neutral particle, which was later called the *neutron*. It had a mass approximately equal to that of the proton. Within a short time the German physicist Werner Heisenberg suggested the nucleus of the atom was made up of protons and neutrons.

Radioactive Decay

It was soon shown that if the nucleus was made up of protons and neutrons, only certain numbers of each gave stability. The simplest nucleus was that of the hydrogen atom; it contained one proton. But there were also hydrogen nuclei with a proton and a neutron, and one with a proton and two neutrons. Atoms with the same number of protons but different numbers of neutrons in the nucleus were called isotopes. The nucleus with two protons was helium, and it appeared in two isotopic forms. In one we have one neutron, and in the other we have two neutrons. We usually refer to them as $^{3}_{2}He$ and $^{4}_{2}He$, where the number at the top is the total number of particles in the nucleus, and the number at the bottom is the number of protons. With this notation, radium is written as $^{226}_{88}Ra$, indicating 88 protons and 226 total particles in the nucleus; ordinary uranium is $^{238}_{92}U$, also known as uranium-238.

Earlier we discussed radioactive disintegration in which alpha and beta rays are given off by radioactive elements. Rutherford discovered that when uranium decays it gives off alpha rays, which are helium nuclei. We can represent this as

$$^{238}_{92}U \mapsto {}^{4}_{2}He +$$

an unknown element.

We can easily determine what this unknown element is by doing a little arithmetic. The total number of protons and the total number of particles on the two sides of the equation have to be the same. Subtracting 4 from 238 and subtracting 2 from 92 gives us our new element of 234 total particles, with 90 protons. Looking at the periodic table we find that the only element with this array of particles is thorium, or $^{234}_{90}Th$. But it is well known this is an unstable isotope of thorium and rapidly disintegrates with the emission of a beta particle. It might not seem the emission of a beta particle, which is just an electron, would change the nucleus. As it turns out, however, the emission of an electron from the nucleus gives the same effect as converting a neutron to a proton. So the total number of particles stays the same, but there is an increase of one proton. And

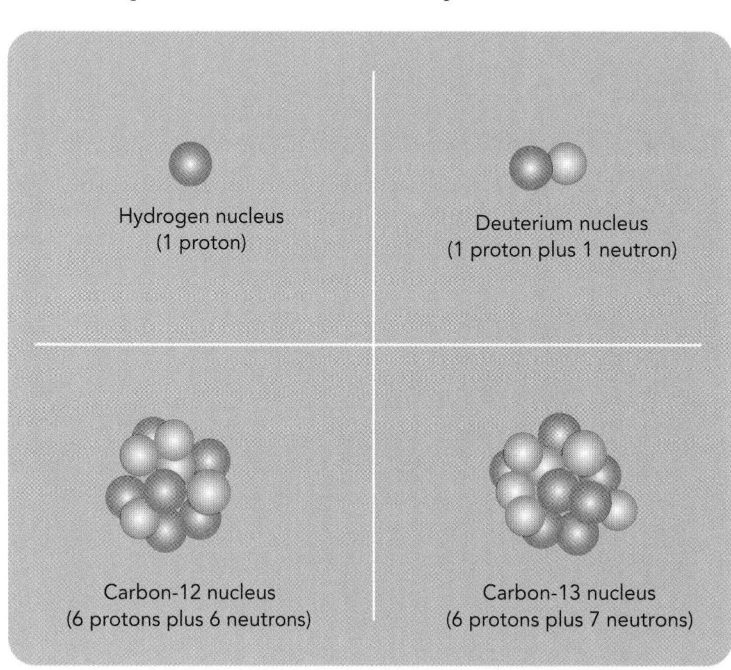

Above: Isotopes are chemically identical variants of an element. This illustration shows the structure of the nuclei of two isotopes each of carbon (bottom) and hydrogen (top). Protons are shown in orange and neutrons in blue.
Top left: The age of coral reefs can be determined using radioactive dating.

in this case we find the element corresponding to the emission of a beta is protactinium, or $^{34}_{91}$Pa.

A SERIES OF REACTIONS

A large number of nuclear reactions exist. In many cases they occur in series. A radioactive element may, for example, decay to another radioactive element that is unstable. After a certain amount of time it will decay to another radioactive element, which may also be unstable, and so on until it reaches a stable nucleus. Some of the better known series are the uranium series, the actinium series, the thorium series, and the neptunium series. In each case the decay takes place through about 10 elements until it finally stops at a stable nucleus. The final stable elements in the uranium, thorium, and actinium series are isotopes of lead. In the neptunium series the final stable element is thallium.

Of particular importance in any reaction is what is called the half-life. It is the time it takes for half of any given amount of a radioactive isotope to decay, leaving the other half. Each radioactive element has its own characteristic half-life. The half-life of uranium ($^{238}_{92}$U), for example, is 4.5 billion years. This means if we have a sample of 20 grams, in 4.5 billion years only 10 grams will remain. All half-lives are not this long, however. The half-life of $^{226}_{88}$Ra is 1,620 years, and the half-life of $^{234}_{90}$Th is only 24.1 days.

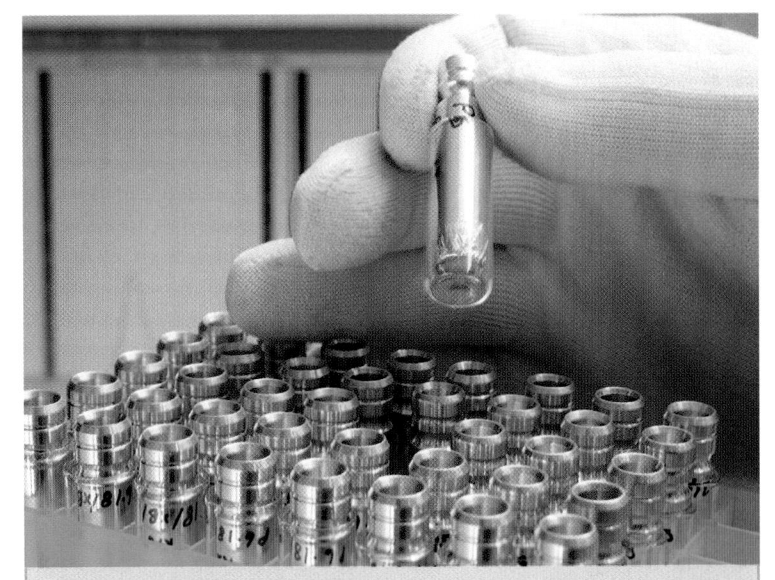

Carbon dating. Vials containing samples for carbon dating using accelerator mass spectrometry (AMS).

RADIOACTIVE DATING

In 1907 the American physicist Bertram Boltwood (1870–1927) suggested that the decay of radioactive elements could be used to determine the age of the Earth's crust. If a layer of rock once contained uranium that had decayed to the stable element lead, a measurement of the rock should give a good estimate of the rock's age. All lead, however, does not result from the decay of uranium, and this caused a problem at first, but it was soon overcome.

Lead has four isotopes, and lead-204 is not produced by decay, so it can be subtracted out. Boltwood's suggestion has turned out to be valuable in estimating the age of the Earth.

Astronaut Charles M. Duke photographed during the Apollo 16 lunar landing mission. The age of rock samples collected during such operations can be determined using radioactive dating techniques.

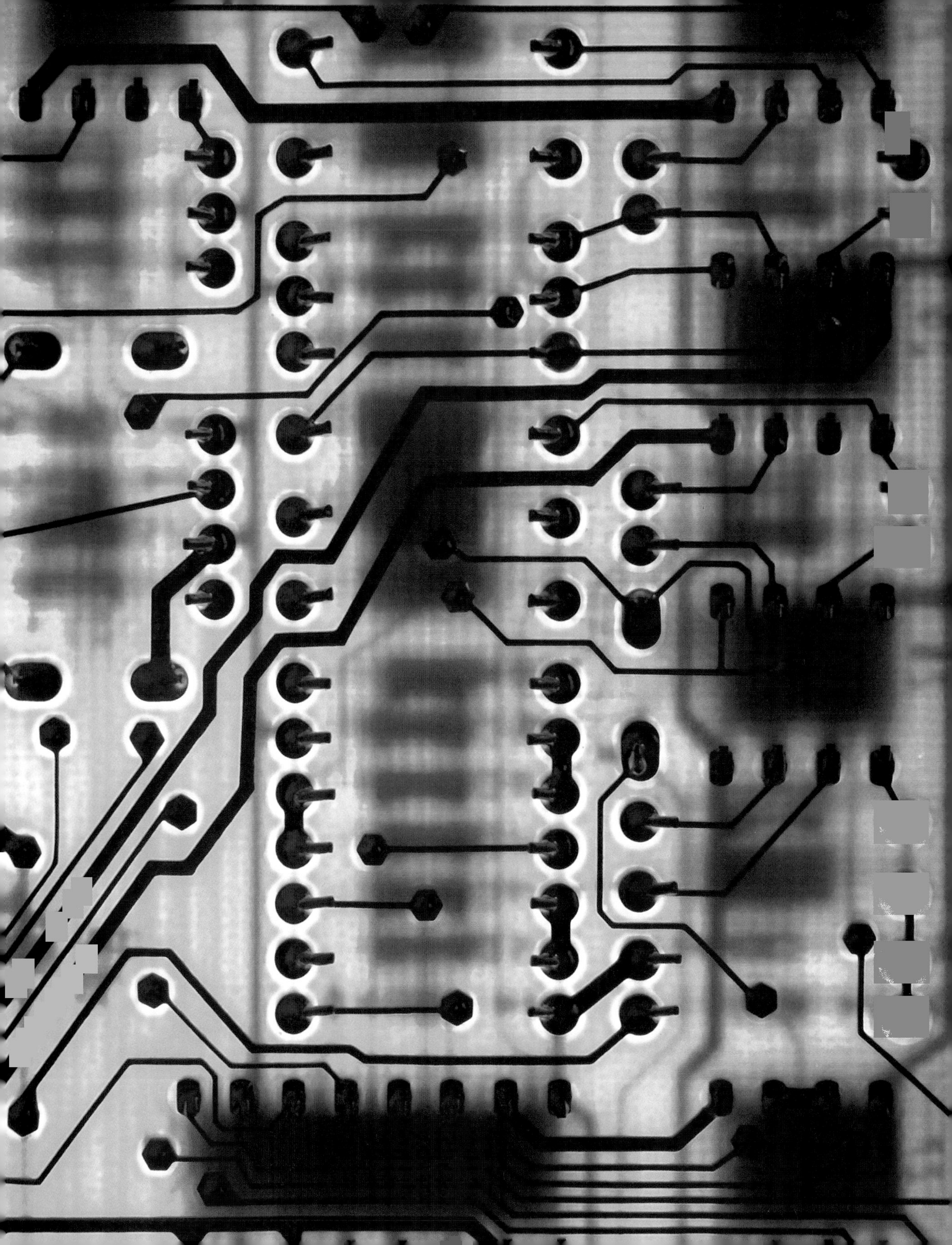

CHAPTER 9

QUANTUM PHYSICS

Left: A computer circuit board. The development of quantum physics has led to the creation of much of the technology that we now take for granted. Top: The first laser was created as a result of experimentation with electron stimulation. Bottom: The study of radiant energy, which causes metal to change color when heated, laid the foundation for quantum mechanics.

Beginning about 1900, researchers and theorists made discoveries that significantly changed physics. The most important of these was the development of quantum theory and quantum mechanics. The word *quantum* in physics refers to the smallest discrete quantity of a physical property. The physicists who developed quantum theory gave us a new and completely different view of the microcosm; in particular, they completely changed our understanding of the atom. It can be said without exaggeration that quantum mechanics is one of the most important scientific developments of all time. And although most people have barely heard of it, it affects the lives of nearly everyone each and every day. Computers, digital watches, television, lasers, and many other modern marvels of technology have come to us as a result of quantum theory.

One of the major problems in physics in the late 1800s was explaining the emission of radiant energy from a glowing hot metal. When a metal was heated it changed color, from red to yellow, then to blue, and finally to blue-white. Physicists tried to explain the phenomenon, but there was a problem: The radiant output depended on the particular metal that was heated. Gustav Kirchhoff (1824–87) of the University of Heidelberg came up with a way to get around this and laid the foundation for the great leaps to come in the field of quantum mechanics.

$h = 6.6262 \times 10^{-27}$ erg s =
= 6.6262×10^{-34} J s

Early Quantum Theory

Kirchhoff suggested using a body that was capable of absorbing all the electromagnetic radiation falling on it; he called this hypothetical object a blackbody. This container, which was a perfect absorber and emitter of radiation, had inner walls of black and only a tiny opening.

Using this blackbody, plots of intensity versus frequency of the emitted radiation were made. They were bell shaped, with little radiation given off at very high and very low frequencies. Furthermore, when plots were made at different temperatures, the peak shifted to a different frequency. (This is what caused the change in color as the temperature was varied.) Radiation of many different frequencies are given off at any temperature, but most of the photons, which are the smallest units, or quantum, of electromagnetic radiation, are emitted in the region of the peak, and the color corresponding to this frequency is the color we see.

Above: Max Planck. He is considered to be the originator of quantum theory. Top left: The formula representing Planck's constant.

The problem was to explain this curve—in other words, to give a formula that fit it. Two scientists took up the challenge about the same time: Lord Rayleigh (1842–1919) of England, and Wilhelm Wien (1864–1928) of Germany. Both men published formulas, but both formulas were flawed. Rayleigh's fit the curve only at low frequencies, and Wien's fit it only at high frequencies.

Rayleigh showed that, on the basis of classical physics, the intensity of the radiation had to be higher as the frequency became higher. But this was not borne out in experiment. There was a definite cutoff: Above a certain frequency there was no emission. The problem eventually became known as the *ultraviolet catastrophe*, because the cutoff occurred in the ultraviolet.

PLANCK'S THEORY

One of those interested in the problem was Max Planck (1858–1947) of the University of Berlin. He worked on it for several years without success, but finally in late October of 1900 he took a different approach. He decided to guess at a formula, then check to see if it fit the curve. And the technique worked: He arrived at a formula that avoided the ultraviolet catastrophe. The formula had a fair amount of flexibility, in that

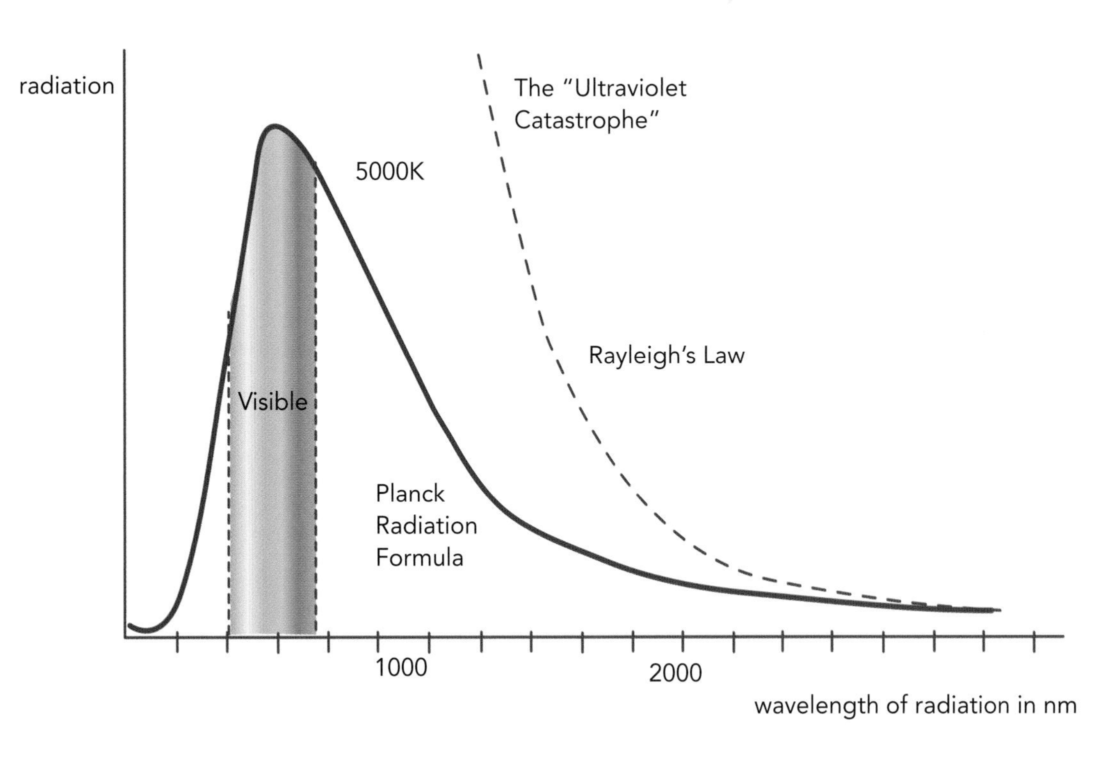

Planck's radiation curve compared to Rayleigh's curve. Experiment confirms that heated objects (blackbodies) obey Planck's formula. The plot is radiation intensity versus wavelength.

it contained two unknown constants that could be determined by comparing to experiment.

He began by assuming the radiation was emitted by tiny "resonators." These were charges attached to entities that oscillated when heat was applied to them. In essence they were forced to accelerate, and according to Maxwell's theory, an accelerating charge gives off electromagnetic waves. One form of this radiation is visible light, and this was the radiation that was of particular interest. What was different about Planck's oscillators was that they did not give off radiation in variable amounts: They gave it off only in discrete amounts. Planck referred to this as "quantization," and he gave the formula for their energy as

$$E = h\boldsymbol{\nu}$$

where $\boldsymbol{\nu}$ is frequency, and h is a constant, now known as Planck's constant; it has the value 6.55×10^{-12} erg-sec.

What Planck did not realize at the time, was that h would play a fundamental part in almost all the physics from that day on. It was a fundamental constant of nature, playing a role similar to the velocity of light.

REACTION TO THE FORMULA

Planck was pleased with the result, but he did not take it seriously. He realized that he had deviated from classical physics in deriving it, and he was sure that it would eventually be explained classically. To him it was just a temporary explanation along the way to a deeper understanding of the problem. And he did not receive a lot of criticism, for almost nobody noticed his breakthrough. In fact, for about five years after the theory was published, it was ignored. Radioactivity and X-rays had been discovered, and they were all the rage. One person, however, did notice it, and his name was Albert Einstein.

Photoelectric Effect and Bohr's Atom

A number of years earlier the photoelectric effect had been discovered by Hertz, but it was not fully appreciated until it was examined in detail by Philipp Lenard (1862–1947) of Germany in 1902. The effect occurs when a light is shone on the surface of certain metals. Lenard showed that electrons are emitted, but not at all frequencies of incident light, and their energy does not depend on the intensity of the light; it depends only on the electrons' frequency. He published his results a few years later.

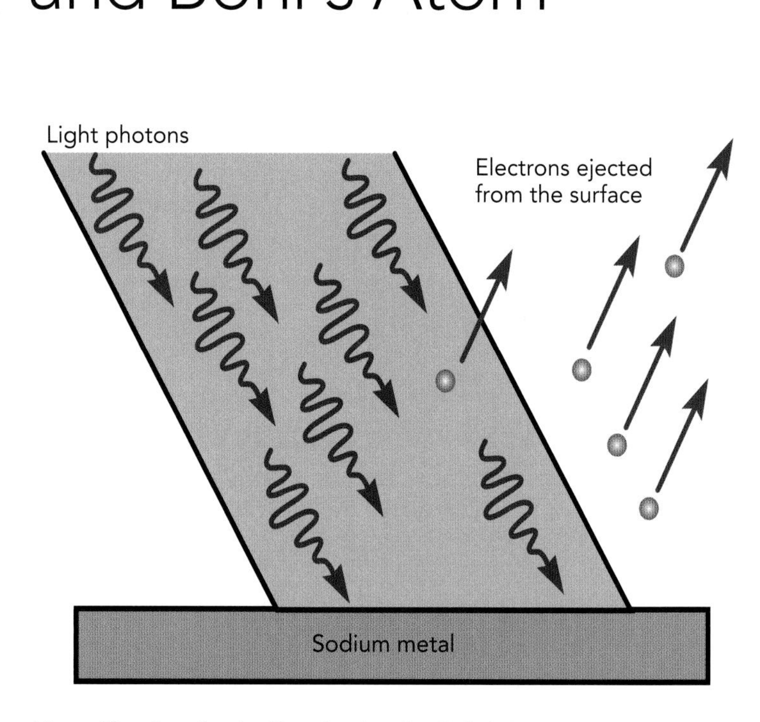

Above: The photoelectric effect showing that light behaves as particles, now called photons. Light strikes the surface of sodium metal in this case, and releases electrons from the surface. Top left: Simple representation of the Bohr model of the atom showing the nucleus and electrons in orbit around it. The electrons have a negative charge and the nucleus has a positive charge.

EINSTEIN'S RADICAL IDEA

Einstein saw Lenard's paper and took an immediate interest in it. He had seen Planck's paper earlier and had been looking for a way to use the "quantum" idea that Planck had introduced. The photoelectric effect seemed to be exactly what he needed. Planck had assumed the emission of the light took place in discrete amounts (quanta), and he had ascribed this to discreteness in the oscillations of the resonators that produced the light. To Einstein, it seemed more logical to assume that light itself was discrete; in other words, it took the form of quanta. Today we refer to these quanta as *photons.* To explain the effect, Einstein assumed that the electron in the surface of the metal absorbed photons of light, and the energy that the electron received depended on the energy of the photon it absorbed. Furthermore, a certain amount of energy was required to pull the electron free from the forces holding it in the surface. If the photon striking

Einstein, photographed around the time that he published his paper on the photoelectric effect (1905). He won the Nobel Prize for it in 1921.

the electron was not energetic enough, the electron would stay in the surface. But if the photon's frequency, and therefore its energy (according to $E = h\nu$), was high enough, it would knock the electron out of the surface.

Einstein submitted the paper for publication in 1905. Coincidentally, it was submitted to Planck, who was the editor of the journal. When Planck looked it over he thought little of it, but he decided to publish it anyway. Years later Einstein was awarded the Nobel Prize for it.

BOHR'S ATOM

Niels Bohr (1885–1962) of Denmark traveled to England soon after he graduated with his doctorate, to work under J. J. Thomson. Thomson had little time for him, however, which was a great disappointment to Bohr. In addition, Thomson assigned Bohr an experimental project that he did not particularly care for, since his main interest was theory. He had not been in England long, however, when Rutherford, who was then at Manchester University, visited Cambridge. Bohr took an immediate liking to Rutherford and soon transferred to Manchester. It was here that he began thinking about the possibility of a "quantum" model of the atom.

LIGHT AS A PARTICLE AND A WAVE

Einstein explained the photoelectric effect by assuming that light was a particle—namely, a photon. The idea was a shock to most scientists. The wave theory of light had been accepted for over a hundred years, and there was a large amount of evidence that light was indeed a wave. So was it a wave or a particle? This was a problem for many years.

Scientists occasionally joked about it by saying that they used the wave theory on Monday, Wednesday, and Friday and the particle theory on Tuesday, Thursday, and Saturday. All joking aside, it eventually became clear that light should not be thought of as *either* a particle or a wave. It is both a particle *and* a wave, depending on how you observe it (or what experiment you are performing). In some cases it appears to be a particle, and in other cases it appears to be a wave. It has a dual nature.

A ray of light. Einstein was the first to propose that light could be thought of as a particle.

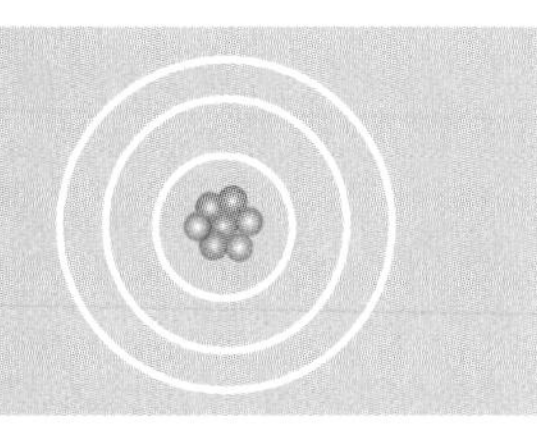

The Quantum and the Atom

Bohr preferred the planetary model of the atom but was familiar with the problem of the electrons spiraling into the nucleus. He knew that a different approach was needed if he was to use the planetary model, and the proposal he made was quite different. Limiting himself to the hydrogen atom, he began by suggesting that the electrons did not emit radiation as they orbited the nucleus. This went against Maxwell's theory and seemed heretical to most scientists. But Bohr was unconcerned; he forged on by assuming that the electrons circled the nuclei in fixed, discrete, orbits, and emitted, or absorbed, energy only when they changed orbits—in other words, when they "jumped" between orbits. Energy was emitted in the form of photons when an electron jumped to a lower orbit; and when it absorbed a photon, it jumped to a higher orbit.

THE BALMER SERIES

The idea that electrons emitted and absorbed photons as they jumped between orbits was radical, but to some it appeared to make sense. Nevertheless, Bohr needed something more to convince the scientific community that it was a valid idea.

He was visiting a friend one day when the friend asked him if his work could explain the Balmer series, which was a series of spectral lines in hydrogen. Bohr, who had paid little attention to spectra and spectroscopy, was unclear about what his friend meant. But the

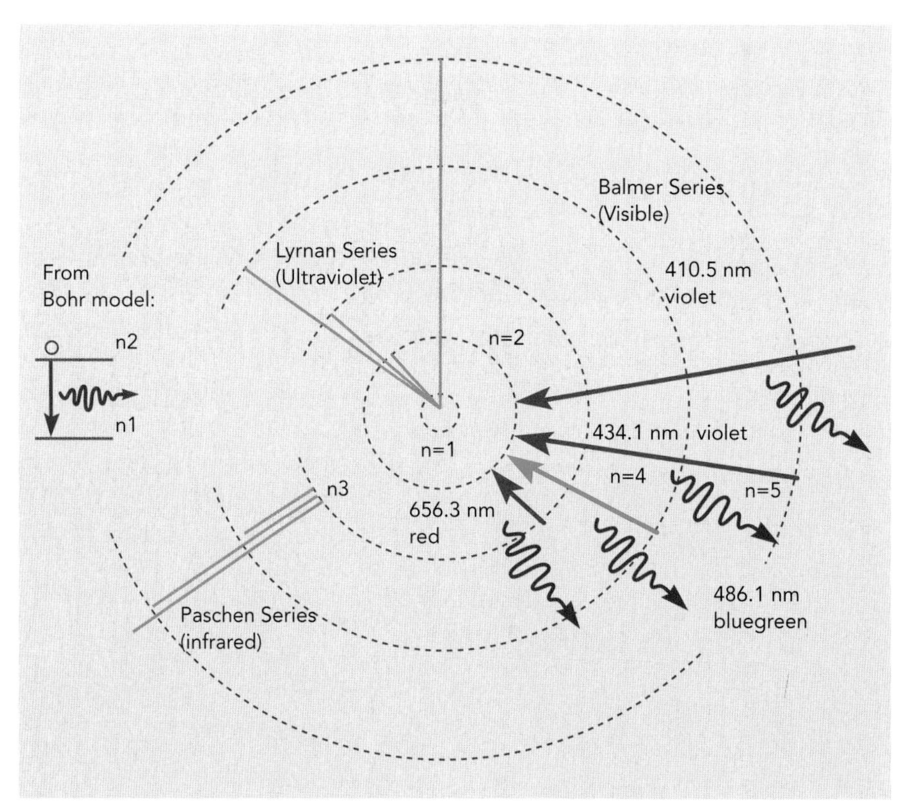

Above: A detailed look at the Bohr model of the atom. The Balmer series is shown on the right; it is a result of electrons falling to level (or orbit) n = 2. Different wavelengths of radiation are given off in each of the cases shown. This radiation is picked up in a spectroscope as the Balmer spectral lines (shown in the lower right). Top left: A simple representation of an atom showing the n = 1, 2, and 3 orbits, where n is the principle quantum number.

Theoretical physicists Arnold Sommerfeld (left) and Niels Bohr. Sommerfeld developed Bohr's ideas on atomic structure.

comment prompted him to go to the library and look into it, and the moment he saw Balmer's spectral formula, he knew it was exactly what he needed.

Although Balmer had given a formula for the spacing of the lines, no one had explained how the lines related to the hydrogen atom. Bohr soon saw that they could be explained using his theory. He derived a formula based on it that related the positions of each of the lines to the orbits; in particular, he showed that the radii of the orbits were 4, 9, 16, and so on, multiplied by the radius of the inner orbit. And of particular importance, Balmer had a constant in his formula, called the Rydberg constant. Bohr showed how it could be calculated in terms of *c*, the speed of light, *e*, the electronic charge, and *h*, Planck's constant. He also gave a formula for the radii of each orbit, along with the energy of the orbit in terms of the same known constants.

REACTION AND VERIFICATION

The initial reaction to Bohr's idea was mostly negative. A number of well-known scientists expressed skepticism, but others soon became strong supporters. Rutherford said little about it. He respected Bohr but found it difficult to accept his strange idea. Within a short time, however, experimental support bolstered the theory. The first came in 1913 from H. G. J. Moseley (1887–1915). He was investigating the emission of X-rays from various substances and showed that the wavelength of the emitted X-rays agreed with the prediction from Bohr's theory. Additional verification came as a result of the work of the German physicist Johannes Stark (1874–1957). He showed that a spectral line split when the emitting atom was in an electric field. A similar effect was found by Pieter Zeeman (1865–1943) for a magnetic field. When Bohr heard of the effects he went to work and soon showed that both could be explained by his theory. Then two German scientists, James Franck (1882–1964) and Gustav Hertz (1887–1975), found that when mercury atoms were bombarded with electrons, they absorbed energy from the electrons. Bohr also explained this result.

Johannes Stark. The German physicist discovered that an electric field would cause splitting of the lines in the spectrum of light, a phenomenon called the Stark effect.

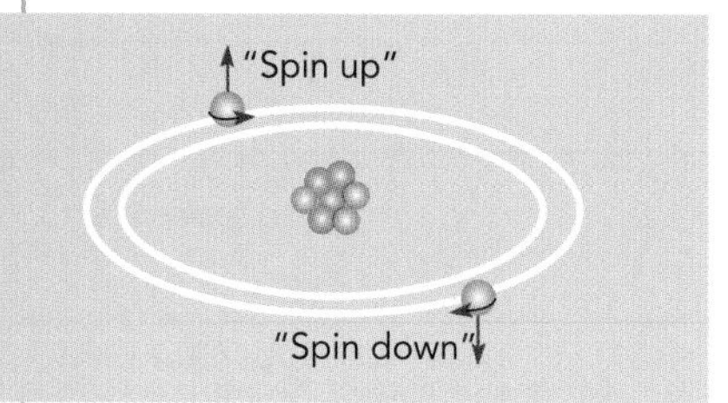

Pauli's Electrons

Bohr's theory on atomic structure was an important contribution to the understanding of the atom, and as it was studied in detail, it became clear that four numbers known as quantum numbers played an important role. The first number, which represented the orbit, was called the *principal quantum number* and was designated by n. It could take on any of the integral values (1, 2, 3, and so on). The second number was introduced by Arnold Sommerfeld (1868–1951). Called the *orbital quantum number*, designated l, it could have any whole number value from zero up to one less than the principal quantum number. Thus, if n was 3, for example, l could be 0, 1, or 2.

After it was found that spectral lines split in a magnetic field, a third quantum number was added; it was called the *magnetic quantum number* and was represented by m. The values of m were the same as l, except they extended into the negative numbers. Thus, if n was 3, m could be any of 0, 1, 2, -1, and -2. Finally, a fourth quantum number, represented by s, was introduced. It was shown that the electron had a *spin,* and this number represented the spin. It could have the values +1/2 and -1/2. In simpler terms, the principle quantum number is the energy of the orbit, the orbital quantum number is the shape of the orbit, the magnetic quantum number is the orientation of the orbit, and the spin is the intrinsic magnetic moment of the electron.

Above: Wolfgang Pauli. The Austrian-born physicist was appointed to the chair of theoretical physics at the Institute for Advanced Study in Princeton and became a naturalized American citizen in 1946. Top left: The electron spins on its axis as it orbits the nucleus.

With these numbers physicists could identify any electron in an atom by its four quantum numbers. But there was a difficulty: It was still not known how many electrons were permitted in each orbit. The Austrian physicist Wolfgang Pauli (1900–58) solved this problem by suggesting that no two electrons in an atom could have the same four quantum numbers. This is known as Pauli's exclusion principle.

A CLOSER LOOK

Consider the first orbit, $n = 1$. According to our rules $l = 0$, m = 0, and s = +1/2, or -1/2. This means that only two electrons can reside in the first orbit, and they have to have opposite spins.

For the second orbit, $n = 2$, we have $l = 0$, 1, and m can take the values 0, 1, -1. More explicitly, if $l = 0$, $m = 0$, and if $l = 1$,

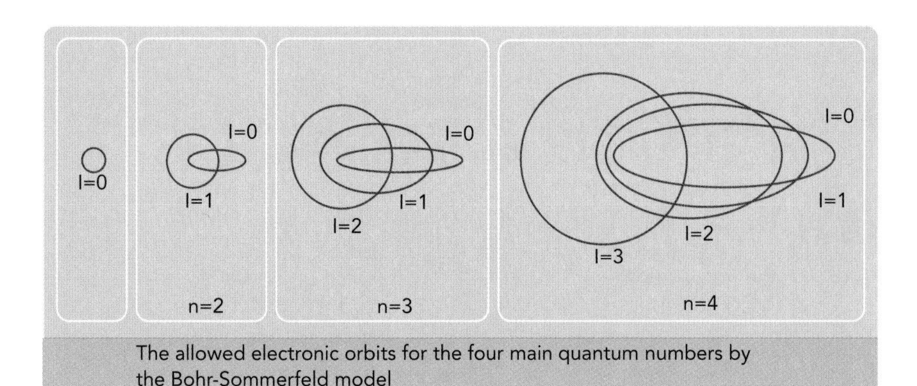

The allowed electronic orbits for the four main quantum numbers by the Bohr-Sommerfeld model

An illustration of the quantum numbers n and l; n represents the orbit number and l is related to the angular momentum of the electron in the state n. For a given n, l can have several values as shown.

then $m = 0$, 1, or -1. This means that for the second orbit we have the quantum numbers [2,0,0], [2,1,0], [2,1,1], and [2,1,-1], and each can have spin number +1/2 and -1/2. In total we therefore have eight electrons in the second orbit. In the same way we can go to the third orbit, where we find 18 electrons are possible. In fact, it can be shown that $2n^2$ electrons may reside in any orbit, where n is the orbit number.

We can take this a step further. The electrons in each orbit can move around the nucleus in any direction, so it is convenient to refer to them as *shells*. They are referred to as the K, L, M, and so on, shells. The electrons within these shells can then be divided into subshells according to their l quantum number. For example, when $n = 1$, we have $l = 0$, and the first electron shell consists of only one subshell, and it contains two electrons. When $n = 2$, however, $l = 0$, 1, and for $l = 0$ we have [2,0,0], and for $l = 1$ we have [2,1,0], [2,1,1], and [2,1,-1], and each can have either +1/2 or -1/2 spin, giving six electrons. This means that the eight electrons in the second shell can be divided into two subshells. It is traditional to designate these subshells as *s*, *p*, *d*, *f*, *g*, *h*, and *i*. We therefore refer to electrons as in the 1s, 2p, subshells, and so on.

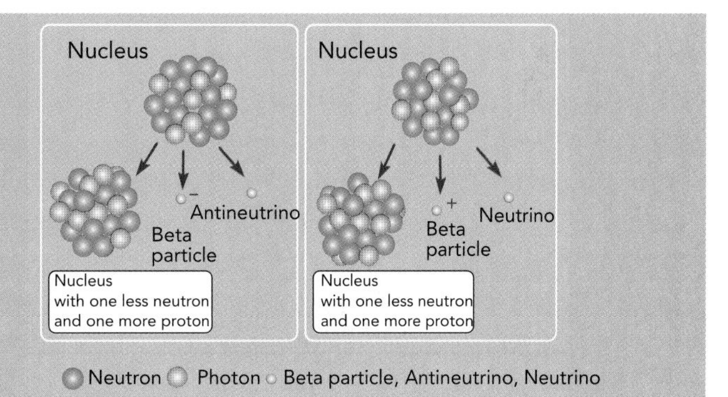

Beta decay can occur in two different ways, as shown above. On the left, a neutron turns into a proton by emitting an anti-neutrino and an electron. On the right a proton turns into a neutron by emitting a neutrino and a positron.

PAULI, BETA DECAY, AND THE NEUTRINO

Several years after formulating the exclusion principle, Pauli made another important contribution to physics. We saw earlier that a neutron in the nucleus can change to a proton with the emission of a beta particle from the nucleus. Similarly, a proton can change to a neutron with the emission of a positive electron. These processes are referred to as beta decay.

There was a serious problem associated with beta decay, however. Such a change involved a definite energy, and the beta particle should have come out with the energy. But this did not occur; beta particles came out of the nucleus with a large range of energies. This was confusing to physicists; it seemed to defy logic.

Pauli solved the problem in 1931 when he suggested that another particle was involved. He predicted that it had no charge and no mass, but it carried away the energy that appeared to be lost. It was called the *neutrino*. With no charge or mass, it was an elusive particle that would be difficult to detect. In fact, it was 25 years before it was finally detected.

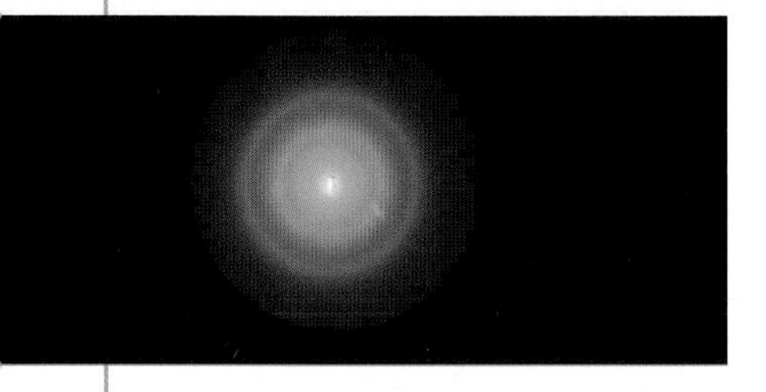

Particles and Waves

The next breakthrough was made in France by the physicist Louis de Broglie (1892–1987). While still in graduate school he developed an interest in waves. A few years earlier Einstein had shown that light was both a wave and a particle; in short, it had a dual nature. De Broglie thought about this in relation to Bohr's orbits. Why did they have the discrete size they had? There had to be a reason. He concluded that if light had a dual nature, it was also possible that the electron had a dual nature. In other words, like the photon, it was both a particle and a wave. And if so, de Broglie was sure that the size of the orbit was related to the wave nature of the electron. He then assumed that the circumference of the Bohr orbits had to be an integral number of the wavelengths associated with the electron. He expressed this as

$$2\pi r/\lambda = n$$

where r is the radius, λ is the wavelength of the wave associated with the electron, and n is an integer.

De Broglie's real insight, however, was to assume that the momentum of the electron (p) was connected with Planck's constant h, and k, which is the wave number, by the relation

$$p = hk$$

These two formulas gave the radius of the orbit in terms of the electron's momentum. It was a "crazy" idea, but de Broglie wrote it up as a doctoral thesis, and in late 1913 he presented it to the faculty of the University of Paris. The idea seemed to make little sense, but de Broglie was from a very important family in France, and his brother was a well-known experimentalist. The reviewers could not just dismiss it. After some discussion it was sent to Einstein for his opinion. "It may look crazy, but it is sound," Einstein replied. Others agreed reluctantly, and de Broglie was awarded his degree.

Above: Louis de Broglie. Top left: A photograph of the diffraction pattern that results when a beam of electrons is fired through a crystalline material.

VERIFICATION

Scientists soon began to look around to see if there was any experimental evidence that particles had waves associated with them. One of the best places to look was in the scattering experiments of electrons from metals, and some of the best experimental results in this

The fifth Solvay conference, Brussels, 1928. Among those present were Einstein, Marie Curie, de Broglie, and Pauli. Belgian physicist Ernest Solvay started these conferences as a forum for experts in the fields of physics, chemistry, and sociology.

area had come from Bell Labs in America. Clinton Davisson (1881–1958) and a colleague, Lester H. Germer (1896–1971), had been scattering electrons from various metals for several years. When Davisson heard of de Broglie's work, he looked at his results and saw nothing that indicated waves were associated with the electrons he was using. De Broglie's prediction, however, was valid only for "single crystal" metals, and Davisson's metals were not single crystal. In April 1925, however, Davisson had an accident in his laboratory that melted some of the metal he was using. When it cooled it became single crystal. Davisson noticed the scattering was changed when he used the metal in a subsequent experiment; furthermore, to his surprise, the results now agreed with de Broglie's prediction. Shortly thereafter George Thomson (1892–1975), son of J. J. Thomson, began a series of experiments in which he projected electrons at thin metallic films. His results also verified de Broglie's results.

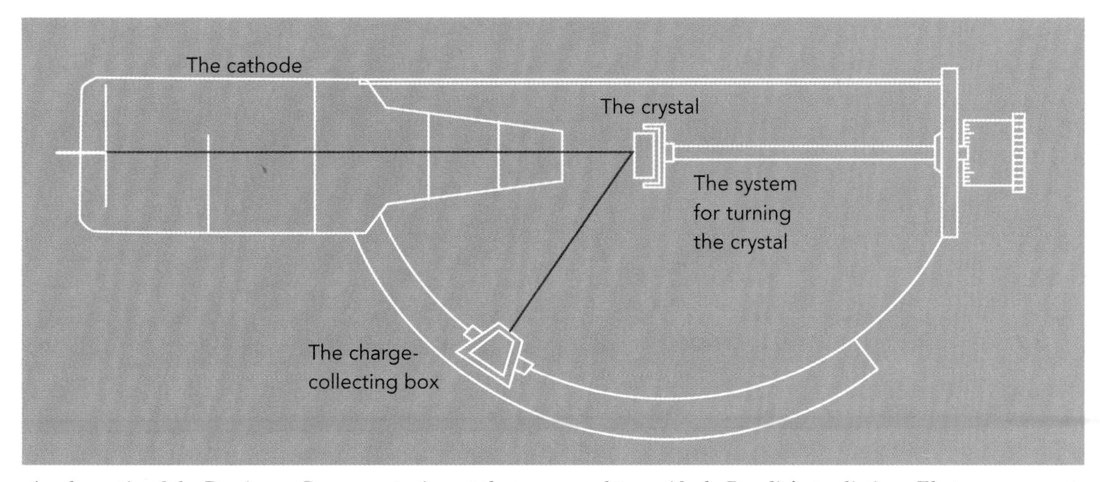

A schematic of the Davisson-Germer experiment that was used to verify de Broglie's prediction. Electrons are scattered from a crystal at various angles. The reflected electrons are collected by the box at the bottom.

Quantum Mechanics and the Atom

Bohr's theory was a first step, but it soon became clear that something more was needed. And it came from the German physicist Werner Heisenberg (1901–76). In early 1925 Heisenberg spent several months struggling with some of the major problems in physics, without much success. As a result, he finally decided to concentrate on the intensity of the spectral lines in hydrogen. They had never been predicted accurately, and he had a new idea on how the problem should be approached. As he started work on it, however, he had an attack of hay fever and decided to go to a tiny island off the coast of Germany called Heligoland to recuperate. He would work on the problem while he was there.

He started by calculating the rates at which electrons jumped back and forth between the various energy levels in the hydrogen atom (called the transition rates). He was dealing with small tables of numbers, and it was convenient to set them up in "arrays." Using these arrays, which unknown to him were known as *matrices* to mathematicians, he calculated the transition rates and was amazed that they agreed exactly with the experimental data.

Heisenberg published his results when he got back to the mainland, and it was soon obvious that he had made a major discovery. His method was referred to as matrix mechanics, or more generally, quantum mechanics. Furthermore, it did not apply just to transition rates; it could be applied to any atomic or molecular problem.

Above: A computer illustration of the "Schrödinger's Cat" thought experiment, which points out one of the famous paradoxes of quantum mechanics. The cat is sealed in a box with a radioactive source and a detector that can detect radioactive particles. It is assumed that the radioactive source will emit a particle with a probability of 50 percent in a period of one minute, and if the particle is detected, a poisonous gas will be released that kills the cat. The radioactive source is turned on for one minute. Is the cat dead or alive at the end of this time? Quantum mechanics cannot tell us for certain; it only gives probabilities. Top left: Werner Heisenberg.

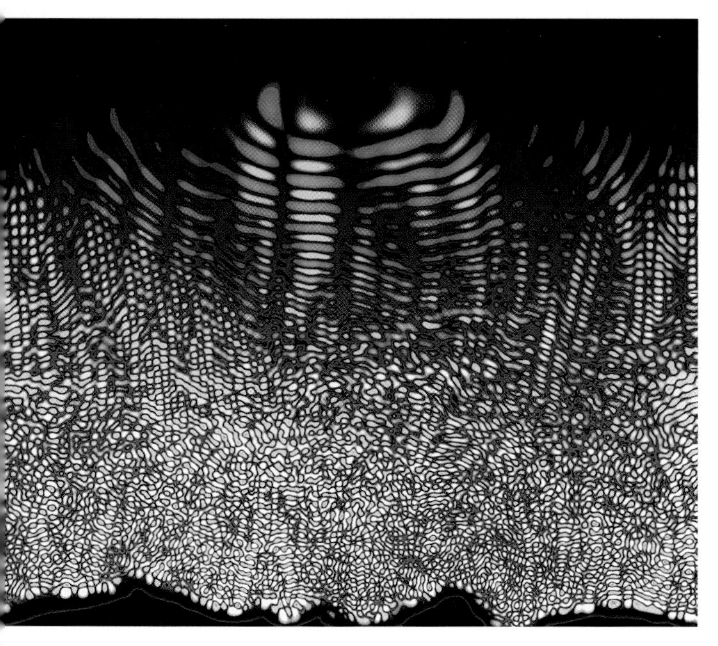

This computer model shows a quantum wave function being reflected from a rough surface.

SCHRÖDINGER'S WAVE EQUATION

Heisenberg's new theory was soon considered to be a significant breakthrough, but most physicists were not familiar with matrices, and as a result, they were not enthusiastic about the new theory. Within a year, however, a completely different approach was discovered by the Austrian physicist Erwin Schrödinger (1887–61). It was based on differential equations, and all physicists were familiar with them, so it was accepted much more readily. Schrödinger was familiar with de Broglie's waves and was sure that they could be described by a wave equation. Such equations had been used extensively in physics to describe other wave phenomena.

He represented the "strength" of the de Broglie wave by a wave function that he called Ψ. After making sure the wave satisfied all the requirements of the Bohr atom, he set up the equation and solved it. The result was an incredibly accurate picture of the standing waves in Bohr's model. He was also able to explain the transitions, or jumps, between the orbits in a new and natural way. Rather than a sudden change, the wave pattern of one state gradually faded as a new one appeared.

Bust of Erwin Schrödinger in a courtyard at the University of Vienna, Austria.

Wave Mechanics and Uncertainty

One of the problems of Schrödinger's new theory, which he referred to as *wave mechanics,* was the wave function Ψ. What exactly did it represent? Schrödinger was unsure. His first paper was published in January 1926, but he did not stop there; sure that he had made a significant breakthrough, he continued to develop the theory. A second paper appeared in February and a third in March. These three papers covered all the major problems in atomic physics, and within a short time the discovery was hailed as an important landmark in physics.

There were, however, two outstanding problems. First of all, how did this theory relate to Heisenberg's theory? Both theories were capable of solving any atomic problem, but they appeared to be entirely different. Schrödinger soon showed, however, that they were equivalent; in other words, Heisenberg's matrices could be constructed from the solutions of Schrödinger's differential equations. They gave the same result.

The second problem was the interpretation of the wave function Ψ. Max Born (1882–1970) of Germany exhibited that it gave a measure of the probability of the position of the electron. More exactly, the square of the wave function expressed the probability that the particle was at a certain point.

THE UNCERTAINTY PRINCIPLE

Heisenberg was not satisfied with the unification of his theory and the wave mechanics that Schrödinger had discovered. In an effort to understand the relationship between them better, he began looking into the details of the measurement process. Four variables—position, momentum, energy, and time—were involved in the process. Heisenberg was convinced that if quantum theory was to make sense, you had to be able to relate the measurement process to observation. In particular, you had to know if it was possible to measure the position and momentum (speed) of an electron at the same time.

> In the sharp formulation of the law of causality—"if we know the present exactly, we can calculate the future"—it is not the conclusion that is wrong but the premise.
>
> —*Heisenberg, in uncertainty principle paper, 1927*

Above: This excerpt is translated from Heisenberg's paper on the uncertainty principle. Top left: The concept of chance, illustrated by a roll of the dice.

Using a "thought experiment," he soon realized this is not possible. Furthermore, he also showed that energy (E) and time (t) cannot be measured simultaneously to a high degree of accuracy. He represented these statements in the formulas

$\Delta x \Delta p \geq h/2$ and $\Delta E \Delta t \geq h/2$.

They are now referred to as the *uncertainty relations*, and the overall principle is called the *uncertainty principle*. What they tell us is that if you focus in sharply on the position (x) of a particle, its momentum (p) becomes fuzzy. Similarly, if you try to focus sharply on p, then x becomes fuzzy ($h/2$ is a measure of this fuzziness). The same goes for E and t.

Niels Bohr's interpretation of the uncertainty principle is named for the city of his birth.

THE COPENHAGEN INTERPRETATION

Heisenberg's uncertainty principle and the probability interpretation that Born introduced eventually led to a "strange" interpretation of quantum mechanics. It was formulated by Bohr of Copenhagen (the reason for the name). The Copenhagen interpretation implied that future events were not necessarily determined by past events. In short, the deterministic view of nature that most physicists had taken for granted up to that time was no longer considered to be valid. The only thing that applied was chance. Many scientists did not agree with this point of view, and it was bitterly contested for several years. Two of the most prominent scientists that were against it were Schrödinger and Einstein. Einstein, along with two colleagues, published a paper in which they gave several arguments for refuting the idea. The paper eventually became known as the EPR (Einstein, Podolsky, and Rosen) paper. In the end, however, Einstein was shown to be wrong.

Lasers

Several modern devices have come out of quantum mechanics. In fact, it is safe to say that our world would be quite different if quantum mechanics had not been discovered. One of these devices is the laser. Lasers are now used extensively in modern society. In medicine, for example, they are used to destroy tumors in the eye, to repair broken blood vessels, and in cataract surgery. They are also used to remove skin lesions, and in the treatment of skin cancer. Lasers are also used in surveying, in cutting metals and other materials, and in several types of welding. The range of applications is extensive.

Above: Schematic of Maiman's ruby laser. At the center is a cylindrical ruby. One end of this ruby is coated with a reflecting mirror, the other end with a partially reflecting mirror. The ruby is surrounded by a coiled flash lamp. Stimulated photons travel along the axis of the ruby, bouncing back and forth as they are reflected. Eventually they break through the partially reflecting mirror and give a laser beam. Top left: Today, the use of lasers is commonplace.

EINSTEIN'S STIMULATED EMISSION

As we saw earlier, electrons can jump back and forth between the various orbits of an atom. When they jump to lower orbits they emit photons of a particular frequency. And if outside energy (heat, for example) is applied to the atoms, electrons jump to higher levels. When in this state, the atom is said to be excited. Einstein considered the case where a photon struck an electron that was already in an excited state and showed that it was possible that the electron could be forced to a lower orbit, rather than jump to a higher one. In this case, two identical photons would be released. He referred to this as *stimulated emission.* When an electron jumps to a lower orbit in the usual way, it is called *spontaneous emission.*

MASERS AND LASERS

Scientists soon found that stimulated emission could be used to create lasers. (The acronym LASER is short for "light amplification by stimulated emission of radiation.") For this, they needed a large number of electrons in an excited state—in essence, a "top-heavy" atom. It is referred to as a *population inversion* and does not occur under normal circumstances in nature. Such an inversion was obtained by a process called "pumping." In effect, the electrons were pumped to a higher energy level. This pumping can be done in several ways. One of the most common is to use an electrical discharge.

Charles Townes (b. 1915) of Columbia University was the first to build a device that created a population inversion, but he did it with microwaves, so his device was called a maser. A few years later, Townes and a colleague, Arthur L. Schawlow (1921–99), were rushing to do the same thing with visible light, but they were beat by Theodore Maiman (b. 1927) of Hughes Research Labs in California. Using a ruby rod and a flash lamp, Maiman built the first laser. In his device, electrons were pumped into excited states by an intense flash from a lamp that was wound around the ruby rod. Upon being stimulated, the resulting photons traveled down the axis of the ruby, bouncing back and forth as they were reflected from mirrors at the inner surfaces at the ends of the rod. As they traveled they stimulated more photons, until the beam was powerful enough to break through the reflecting surface at one end.

Laser levels allow for greater accuracy of measurement.

Maiman's was the first laser, but many other types are used today. Some are pulsed beam and others are continuous beam; the former are more powerful, but the latter are more useful in medical applications.

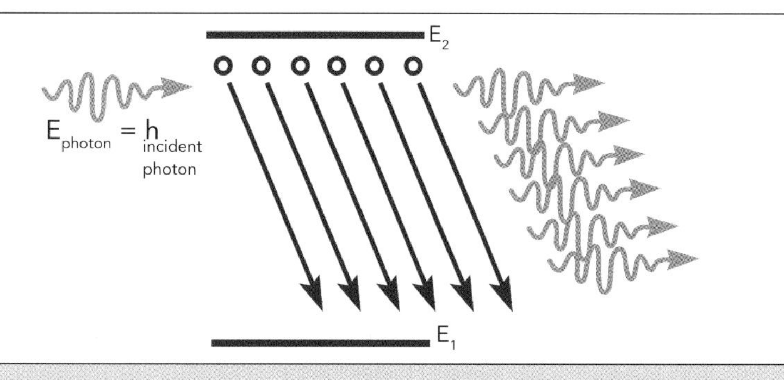

Stimulated emission of electrons from a high energy level to a lower level. The emitted photons will all be in phase—in other words, their waves will all be lined up, or coherent.

COHERENT LIGHT

When electrons are pumped so that they jump to higher energy levels, as we saw, we get a population inversion. This inversion then has to be stimulated to fall to a lower energy level. In this case the photons will all have exactly the same properties. Furthermore, we will get more photons than we put in to stimulate them, so we will have amplification. In addition, all the photons will have the same frequency, and their waveforms will be *coherent*. What this means is that they will all be lined up as in the figure. In ordinary light they are not lined up, and this causes "internal scattering." Internal scattering depletes photons from the beam, so it cannot be made very intense, or at least it will not remain intense for long. If, for example, you shone a large searchlight into space, the beam would not get far before internal scattering weakened it. A laser beam, on the other hand, can be shone all the way to the Moon. When the first astronauts landed there in 1969, they reflected a laser beam from a small mirror they had set up on the Moon's surface.

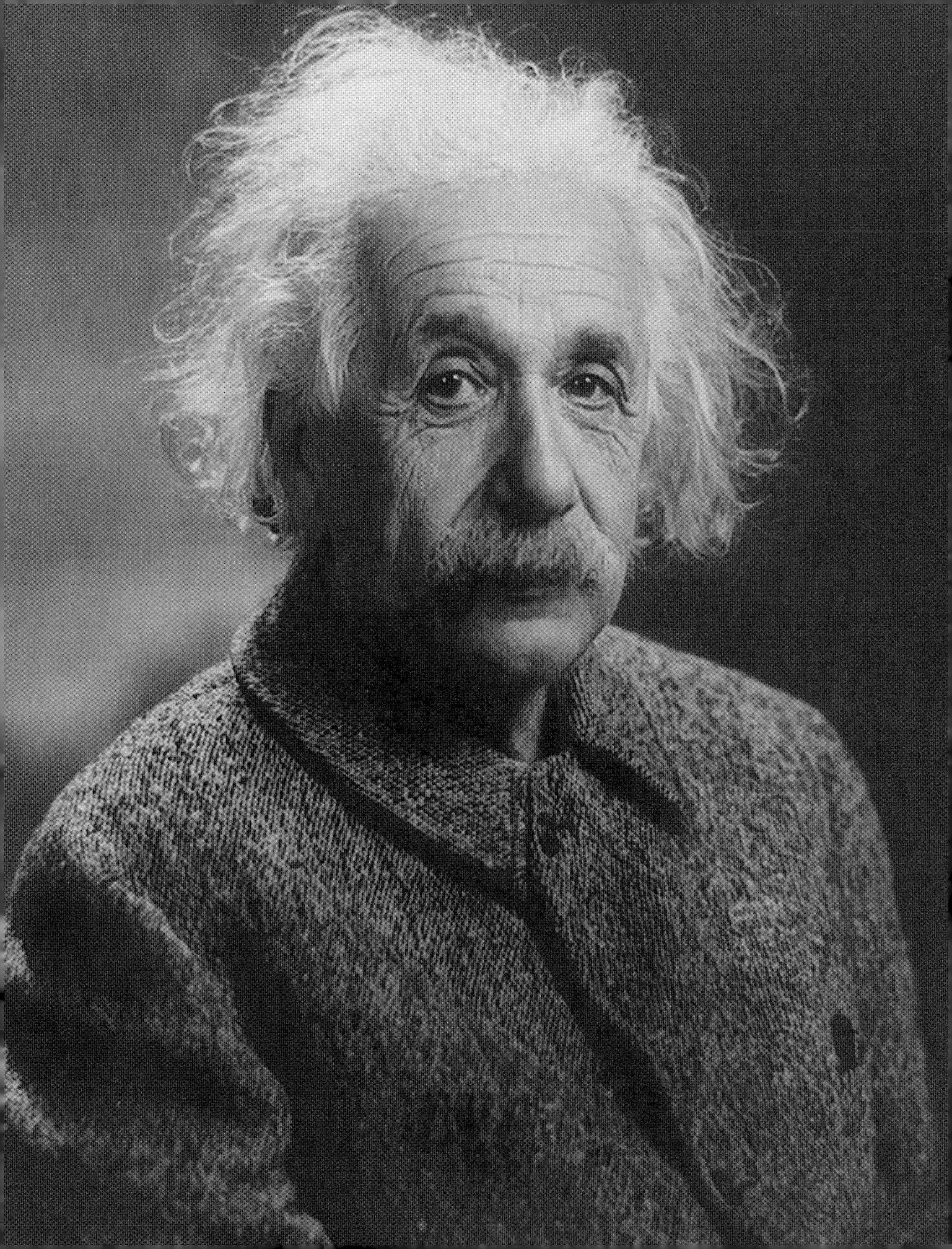

CHAPTER 10

EINSTEIN AND RELATIVITY

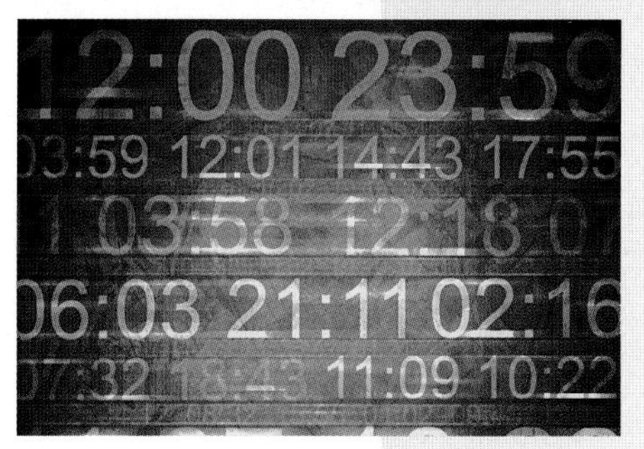

Left: Albert Einstein in 1947 at the age of 68. Top: Einstein's famous equation, $E = mc^2$, showed that energy and mass are equivalent. Bottom: Time montage. Time is a central concept in relativity.

The theory of relativity consists of two separate theories referred to as special relativity and general relativity. Both theories introduced a number of strange concepts into physics. The special theory deals with space and time and shows that they are much different from what most scientists once assumed. According to Newton, space and time were absolute—never changing, always the same. Einstein challenged him and showed that neither is absolute; both are flexible and depend on how you see them.

General relativity also introduced strange new concepts, one of the strangest of which was curved space. According to Einstein's theory, gravity is curved space—a concept that few people understand, and literally no one can visualize. Yet it proved to be a better theory than Newton's; it correctly predicted several things that Newton's theory did not.

Einstein, the Man

Albert Einstein was born in Ulm, Germany, in 1879 but lived there only a year. In 1880 his family moved to Munich, where he attended school until the equivalent of our tenth grade. Few teachers recognized his talent. In 1894 his father faced business failure and decided to move the family to Milan, Italy. To Einstein's dismay they left him in Munich to finish high school. Homesick, he soon worked out a scheme to join them, obtaining a letter from his doctor saying he was on the verge of a nervous breakdown.

Einstein at about age 14. His family moved from Germany to Italy around this time. Einstein received his high school diploma at the age of 17.

EINSTEIN AT SCHOOL

For the next year Einstein thoroughly enjoyed himself, traveling around Milan and throughout Italy, but he had promised his parents that he would take the entrance exams for the Polytechnic in Zurich, so he eventually had to buckle down and study. In 1895 he took the exams and failed. Although his grades in physics and mathematics were high, he had neglected to study most other subjects, so his overall grade was too low for him to enter the university. The officials suggested that he finish high school in the nearby town of Aarau, and a year later Einstein had his diploma. At the age of 17, he entered Zurich Polytechnic in a class of five students: three math majors and two physics majors. The other physics major was a girl, Mileva Maric (1875–1948), whom Einstein eventually married.

Einstein was not a particularly conscientious student and spent most of his time in the laboratory. He fell behind in his studies and had to cram for the finals. Although Einstein had placed at the top of his class in the midterm, he placed fourth out of five in the finals; he beat only Maric, who failed. The next couple of years were a low point in his life. He had hoped to get an assistantship with one of his professors, but no one would hire him.

Above: Einstein in the 1920s. By this time he had become world famous, as his general theory of relativity had been proven correct in November 1919. Top left: Einstein and his wife, Elsa.

EARLY CAREER

Finally he got a job at the patent office in Bern, where he and three friends began what they called the "Olympic Academy."

Einstein was supposed to be their tutor, but the meetings were more of a mutual exchange of ideas, with Einstein leading the group. During the years that he was involved with the academy, Einstein pursued an extensive self-learning program and began looking into many of the fundamental problems of physics.

By 1905 he was hard at work on several of the most difficult problems in physics, and published five of the most important papers known in the field. One of them was his special theory of relativity, a paper that completely changed established views of space and time. Another was on quantum theory and the photoelectric effect—a paper that eventually won him the Nobel Prize. A third paper was submitted as a thesis to the University of Zurich, for which he was awarded a doctorate.

Einstein became an assistant professor at the University of Zurich in 1911. He held positions at the University of Prague and the Polytechnic in Zurich before settling at the University of Berlin's Wilhelm Institute for Physics.

In 1919, Einstein's marriage, which had produced three children, was dissolved. In the same year he married his cousin, Elsa Löwenthal. As the Nazis rose to power, Einstein began speaking out against Hitler, and it soon got him in hot water, so he left Germany in 1933 and went to the Institute for Advanced Study in Princeton, New Jersey, where he remained until his death in 1955.

Einstein, an ardent pacifist, renounced his German citizenship after Hitler became chancellor in 1933. He moved to Princeton, New Jersey, where he worked at the Institute for Advanced Study, played violin, and enjoyed the home that he called "a wonderful little spot."

EINSTEIN, THE LEGEND

Albert Einstein is considered by many to have been the greatest scientist of the twentieth century. Many stories have been told about him, a number of which have become legend. Here are a few such tales:

Einstein's secretary at Princeton received a telephone call, in which someone with a strong German accent asked for Dr. Einstein's address. She said she was sorry, but she had strict orders not to give out the information. "That's a problem," said the caller. "You see, I am Dr. Einstein and I have forgotten my address, and I would like to come home."

Einstein posed for photographers, artists, and sculptors so often that when he was asked by a stranger one day what he did for a living, he replied, "I'm an artist's model."

Einstein was sitting next to an 18-year-old girl at a party when she asked him what he did. "I study physics," he replied. She looked at him strangely. "At your age? . . . I finished my physics over a year ago."

When Einstein received a telegram telling him that his prediction of a shift in position of the stars near the limb of the Sun during an eclipse had been verified, a student asked him what he would have done if it hadn't been verified. "In that case I would have felt sorry for the dear Lord, because the theory is correct."

Special Relativity

If light was a wave, it had to be like other waves and therefore needed a propagating medium. Water waves, for example, need water to propagate them; without the water there would be no waves. The propagating medium for light was not obvious, so scientists invented one; they called it the aether. This aether presumably permeated all space and possessed some amazing properties: It had to be transparent and rigid, but it could not resist or restrict matter in any way, and it had to be unaffected by gravity. The concept of the aether gave rise to a fixed frame of reference for the universe. This meant that the Earth, for example, had a velocity relative to the aether. Even if it was possible that the Sun was at rest relative to the aether, the Earth was traveling around the Sun, so the Earth had to be moving relative to it.

THE MICHELSON-MORLEY EXPERIMENT

In the early 1880s the American physicists Albert Michelson and Edward Morley set out to confirm the existence of aether by measuring Earth's velocity through it. The easiest way to do this was to measure the velocity of a light beam. The aether carried the light beam, so the velocity of light relative to it had to be constant. But the Earth was moving around the

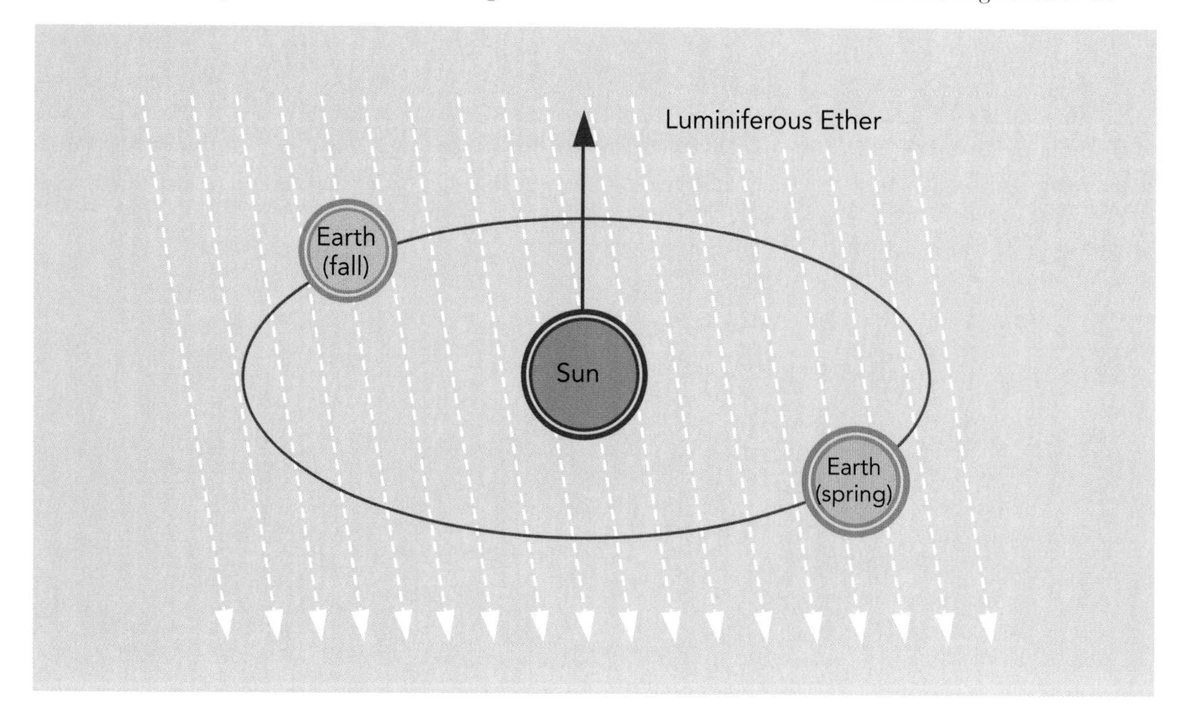

Above: The Sun with the Earth orbiting it. Surrounding them is the hypothetical aether. It was assumed early on that the Earth was moving through this aether. Top left: Einstein at the Huntington Library in California. At left front is Albert Michelson; on the right is Robert Millikan.

Sun at 29 km/sec (its orbital velocity), so if a beam of light was projected in the direction of the Earth's motion, it should appear to be going $c - 29$ km/sec, where c is the velocity of light. If the beam was projected in the opposite direction, it should appear to be going $c + 29$ km/sec.

Michelson and Morley set out to measure this difference and were surprised by the result. Regardless of which direction the beam was projected (with or against the Earth's motion), the velocity of light was always c. In other words, the motion of the Earth had no effect on the speed of the light beam. What this meant was that you could not catch up with a light beam; regardless of how fast you went, it would always travel c faster than you.

Scientists were stunned. The result did not seem to make sense. The only explanation was that objects in motion shrunk in the direction of their motion. But few believed this was possible. Two scientists, H. A. Lorentz (1853–1928) of Holland and G. F. Fitzgerald (1851–1901) of Ireland, derived a formula that gave the amount of shrinkage; but there was no explanation of why it occurred. Their formula can be written as

$$L = L_0 \Upsilon$$

where L_0 is the original length of the object, L is the length we observe, and $\Upsilon = (1- v^2/c^2)^{1/2}$, where v is the velocity of the object relative to us.

EINSTEIN'S THEORY

Several years after the Michelson-Morley experiment, Einstein considered the problem of what was invariant, or the same, for all observers. He was interested in an enigma that occurred in relation to electric and magnetic fields. In 1905 he published a paper on the observed constancy of light speed, without assuming the presence of aether. Einstein set out two postulates:

1. The laws of physics are the same in all uniformly moving systems.
2. The speed of light is the same regardless of the motion of the source.

So, all motion was relative; there was no such thing as absolute motion. Einstein was able to derive the formula that Lorentz and Fitzgerald had derived earlier, but he now had an explanation of it.

The theory of relativity predicted several other things. For example, it showed that time slows down relative to a fixed observer as you approach the speed of light, and it showed that mass also increases as you approach this speed. There are, in fact, changes at all speeds, but they are significant only near the speed of light.

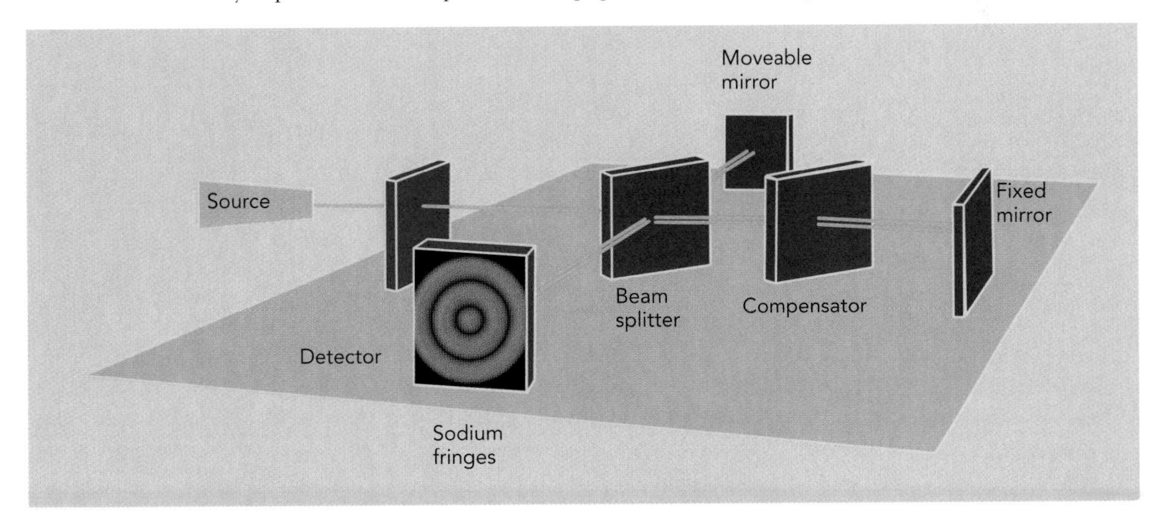

This diagram represents the Michelson interferometer that was used in the Michelson-Morley experiment. A light beam was split so that half of it went to one mirror (called the moveable mirror) and half to another (called the fixed mirror). The reflected rays were brought back together and caused an interference pattern at the detector.

Velocity of Light and the Twin Paradox

Looking closely at Einstein's theory, we see that not only do things shrink in the direction of motion, time slows down, and masses increase as velocities get very high, but at the speed of light things get even stranger. At the speed of light all objects shrink to nothing, time stops, and masses become infinite. Since these things cannot occur in reality, then the speed of light must be unattainable. We now know that only light particles can travel at the speed of light; matter can travel at any speed up to the speed of light, but not *at* the speed of light.

Below: This illustration depicts a spaceship traveling at nearly the speed of light. Top left: A colorful display of light rays.

THE TWIN PARADOX

After Einstein showed that time slows down at high velocities, it was obvious that space flight to the stars might be possible. If a spaceship traveled at an exceedingly high velocity (of the order of 0.999*c*), you could travel to a nearby star and back in a relatively short time (according to your watch)—perhaps a few months. But according to Einstein's formula, time would pass differently back on Earth while you were away; in fact, a large amount of time would pass. So when you returned to Earth a month or so later (according to your watch), many years would have passed—perhaps hundreds of years.

And even stranger than that is the scenario of the twins. Assume a pair of twins are both age 25 on Earth. Assume further that twin A gets in a spaceship and travels to a star at high speed. When he gets back he will be only 25 plus one month older, whereas his twin might be 75 or more. But according to relativity we can look at this another way. Einstein said that all motion is relative, so it is equally correct to say that the twin in the spaceship stood still, while the Earth moved off into space at high speed and came back. In this case, twin B would be the younger. This was obviously a problem, and it was solved when Einstein published his general theory of relativity. The theory told us that one twin would experience forces due to the acceleration, and that he would be the younger of the two.

An artist's interpretation of the phenomenon of time slowing down as it does at high relative velocities.

THE MOST FAMOUS FORMULA IN THE WORLD

Not only did Einstein's theory clear up many mysteries about space and time, it also cleared up one of the major mysteries in relation to stars (including our Sun): Where did they get all their energy? Using his new theory, Einstein derived a formula for energy, and what he got surprised everyone. It told us that mass and energy are equivalent according to

$$E = mc^2$$

where E is energy, m is mass, and c is the speed of light. In short, if we could convert mass directly to energy, the amount of energy we would get is given by the above formula. Since c is extremely large (299,792 km/sec), and its square is even larger, the amount of energy that we would get from even a small amount of mass is obviously incredible. This is the source of energy of the stars, and it is also the source of energy in the atomic and hydrogen bombs.

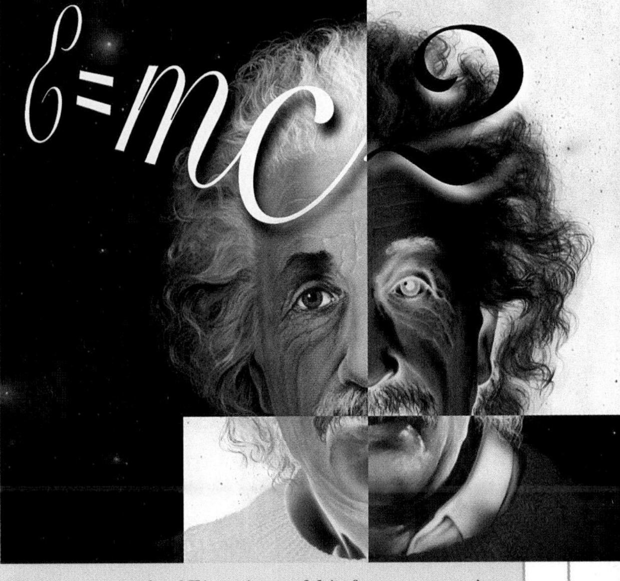

Abstract artwork of Einstein and his famous equation.

General Relativity

Einstein's theory of special relativity, which was published in 1905, applied only to straight-line, uniform motion. Einstein was therefore soon hard at work trying to extend it to accelerated motion. This was a much more complicated problem, and it took him 10 years to solve it. Again, as in special relativity, he started with two postulates: first, that no observer, regardless of how he is moving, can determine his state of motion by experiment; and second, that gravitation and inertia are equivalent. Inertia is the force we feel when we are accelerated. When a car accelerates, for example, you are forced back against the seat by an inertial force. The second of the above postulates is usually referred to as the equivalence principle.

THE EQUIVALENCE PRINCIPLE

To understand some of the implications of the equivalence principle, consider an observer in an elevator near the top of a high building. Assume the cable holding the elevator breaks and the elevator falls. If the observer in the elevator jumps, he would float up from the floor, as if he were in space. Furthermore, if he tried to drop an object, it would stay where he released it (it would not fall to the floor).

Now, assume we take the elevator into space and accelerate it upward with an acceleration equivalent to the acceleration of gravity on Earth (9.8 m/sec^2 or 32 ft/sec^2). According to the equivalence principle, everything in the elevator should be the same as it is on Earth, and the observer will therefore assume he is on Earth. Now, suppose he drills a hole in one side of the

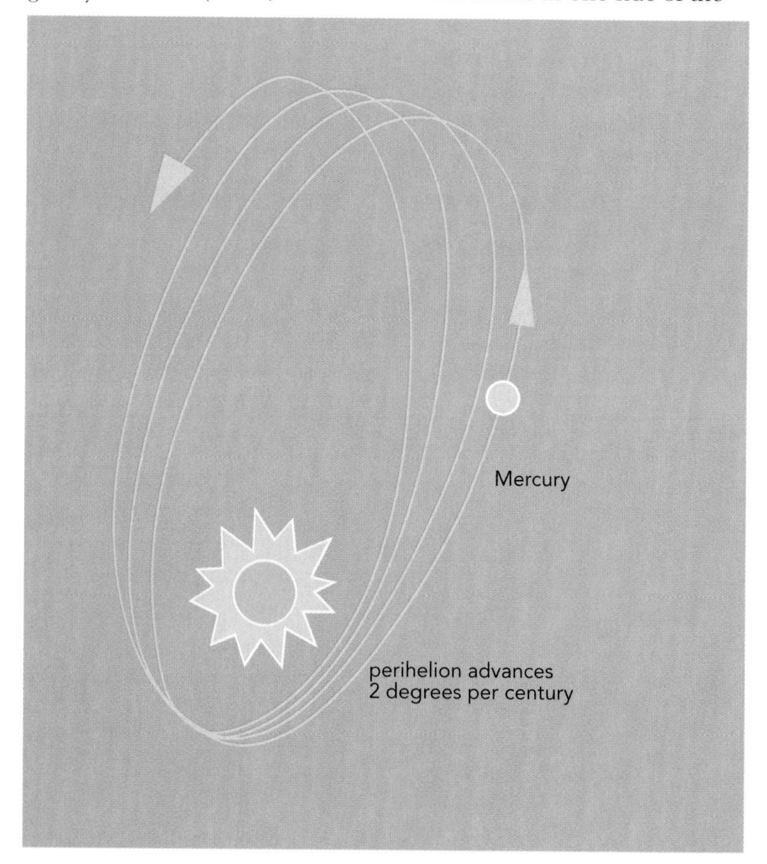

Above: The precession of the orbit of Mercury. The axis of the above ellipse changes position slowly over the years according to general relativity. Einstein predicted a precession of 43 seconds of arc per century. Top left: An elevator shaft. Many of Einstein's so-called thought experiments took place in an elevator. In one such experiment, it was shown that a light beam would bend as it passed through a gravitational field.

elevator and allows a beam of light to pass through the interior of the elevator. Since the elevator is accelerating upward, the beam will be deflected downward as it crosses the elevator. But again, according to the principle of equivalence, we should expect the same thing to occur in a gravitational field. In other words, a gravitational field should deflect a light beam, and as we will see, this is easy to test.

GRAVITY AND CURVED SPACE

Einstein's general theory of relativity, which was published in 1916, was a remarkable theory. Like Newton's theory of 300 years earlier, it was a theory of gravity, but it interpreted gravity in a completely different way. Where Newton assumed it was an action-at-a-distance force, Einstein assumed it was curved space. Since the space was curved, Einstein could not use ordinary Euclidean geometry to describe it. He had to use a non-Euclidean geometry that had been devised by the German mathematician Bernhard Riemann. It is a geometry of positively curved space, like the surface of a ball. One way to characterize geometry is to consider the sum of the interior angles in a triangle. In ordinary flat space, where we use Euclidean geometry, the sum of the interior angles is 180 degrees. But for a positively curved surface, such as the surface of the Earth, this sum is greater than 180 degrees. In a third geometry, devised by the Russian Nikolai Lobachevsky (1792–1856), the sum is less than 180 degrees; it is said to be negatively curved.

Einstein used non-Euclidean geometry to develop his theory of curved space. He assumed that matter curved space, and that when other matter encountered this curvature, it moved through it in a "natural manner." We refer to this natural manner as following a curve called a geodesic (a geodesic is the shortest distance between two points). This path would appear curved to us because the space is curved. This means our Sun curves the space around it, and all the planets, including Earth, move in geodesics in this curved space.

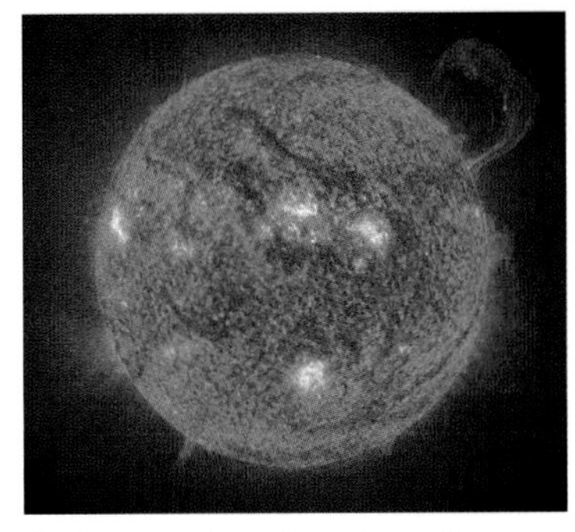

A close-up of the Sun. According to Einstein's theory, the Sun curves the space around it and all the planets, including Earth, move in geodesics in this curved space.

TESTING THE THEORY

One of the things any theory has to do is make predictions that can be tested, and Einstein pointed out several tests that could be made on general relativity. The first is to show that a beam of light that passes close to the limb of the Sun is deflected. This deflection was verified by Sir Arthur Eddington in 1919, and has been verified many times since. A second test involves the orbit of Mercury. According to both Newton's and Einstein's theories, the planets should orbit the Sun in elliptical orbits. Newton's theory predicted that the axis of the ellipse should *precess*, or change its direction slowly, by an amount smaller than what is actually seen. Observations have shown that Mercury's orbit does precess almost exactly according to Einstein's prediction (43 seconds of arc per century).

A third prediction made by Einstein's theory is that time slows down as the gravitational field increases. This means that someone living on the top story of a skyscraper ages faster than someone living on the first floor (the difference, however, is exceedingly small). This has been tested and has been shown to be correct.

One of Einstein's predictions was that a clock runs slow in an increased gravitational field. A clock at the top of this building would therefore run faster than one on the first floor.

CHAPTER 11

FISSION, FUSION, AND THE BOMB

Left: The blast of a hydrogen bomb. Top: A nuclear reactor power station. Bottom: Particle physicists (left to right) Niels Bohr, Hideki Yukawa and his wife, and Robert Oppenheimer. Bohr formulated the first quantum model of the atom, Yukawa predicted the existence of mesons, and Oppenheimer led the U.S. team that developed the atomic bomb in Los Alamos, New Mexico.

In the years after Einstein showed that energy and matter were equivalent, there was considerable debate and speculation. Was it possible that matter could be transformed to energy? And how would we go about it? Einstein and many other well-known scientists were skeptical. When asked about the possibility, Einstein replied, "There is not the slightest indication that this energy will ever be obtainable. It would mean that the atom would have to be shattered at will. . . . We see atomic disintegration only when nature herself presents it."

Occasionally, it seems, Einstein was wrong. But tapping the new energy source would not be easy. The key discovery came in 1939 when two German chemists, Otto Hahn (1879–1960) and Fritz Strassmann (1902–80), accidentally discovered something strange when they bombarded uranium with neutrons. Within months scientists realized that the magic wand was a process called fission. Through fission, an unfathomable abundance of energy was possible. The energy release was so great it could be used to supply the needs of mankind for years, but, alas, it also could be used to build a bomb. About the same time another even greater source of energy, called fusion, was discovered. With these two discoveries the world had entered the atomic age.

A More Stable Nucleus

By the 1920s it was well established that the electrons in the atom were held to the nucleus by the electrostatic force. But what held together the nucleons (protons and neutrons) in the nucleus? They did not attract one another electrostatically; in fact, there would be a large repulsive force between the protons. Whatever this binding force was, it had to be much stronger than this electrostatic repulsive force. Furthermore, it could not affect the electrons that were whirling around the nucleus, so it had to be short-ranged, which would make it quite different from the electrostatic force. Also, at least three separate forces were needed: one to hold the protons together, one to hold the neutrons together, and one to hold the neutrons to the protons.

The discovery of quantum mechanics dramatically changed accepted notions of how the electrostatic force operated. From a quantum mechanical point of view, it was caused by an exchange of particles, in particular, photons. A negatively charged electron, for example, was attracted to a positively charged proton as a result of the exchange of photons between the two particles.

In 1935 the Japanese physicist Hideki Yukawa (1907–81) decided to extend this exchange idea to the nucleus. He suggested that the nucleus was held together by particles moving

MESONS $q\bar{q}$ Mesons are bosonic hadrons There are about 140 types of mesons.					
Symbol	Name	Quark content	Electric charge	Mass $G = V/c^2$	Spin
¶$^+$	plon	ud	+1	0.140	0
K$^-$	kaon	su	-1	0.494	0
ρ+	rho	ud	+1	0.770	1
B^O	B-zero	db	0	5.279	0
η_c	cta-c	cc	0	2.980	0

Above: A meson is a particle made from a quark-antiquark pair. This chart gives five examples of the 140 known mesons. The quark content shown reflects this pairing. Top left: A simple representation of an atom.

back and forth between the nucleons, a force he called *strong nuclear force*. Because it had to be very short-ranged, the exchange particle of the nucleus had to have mass. From the range of its field Yukawa was able to calculate that its mass should be about 200 times that of the electron. Yukawa also knew that he would have to account for the attraction between both types of nucleons, and this meant that the exchange particle between a proton and a neutron had to be charged. When it was absorbed by a neutron, for example, the neutron would become a proton, and when a proton emitted an exchange particle, it would become a neutron.

Yukawa published his theory in 1935. The particle that he predicted was intermediate in mass between the electron and the proton (or neutron), and therefore it was called a *meson*. In 1936 the American physicist Carl David Anderson (1905–91) discovered such a particle. It had a mass 130 times that of the electron, and was called a mu-meson, or muon. But it was soon shown that it did not interact with the nucleus, so it could not be Yukawa's particle. In 1947, however, another meson was discovered by the English physicist Cecil Frank Powell (1903–69); it was called the pi-meson, or pion, and it had all the properties of Yukawa's particle. Yukawa was awarded the Nobel Prize in 1949 for his prediction. Anderson and Powell were also Nobel laureates.

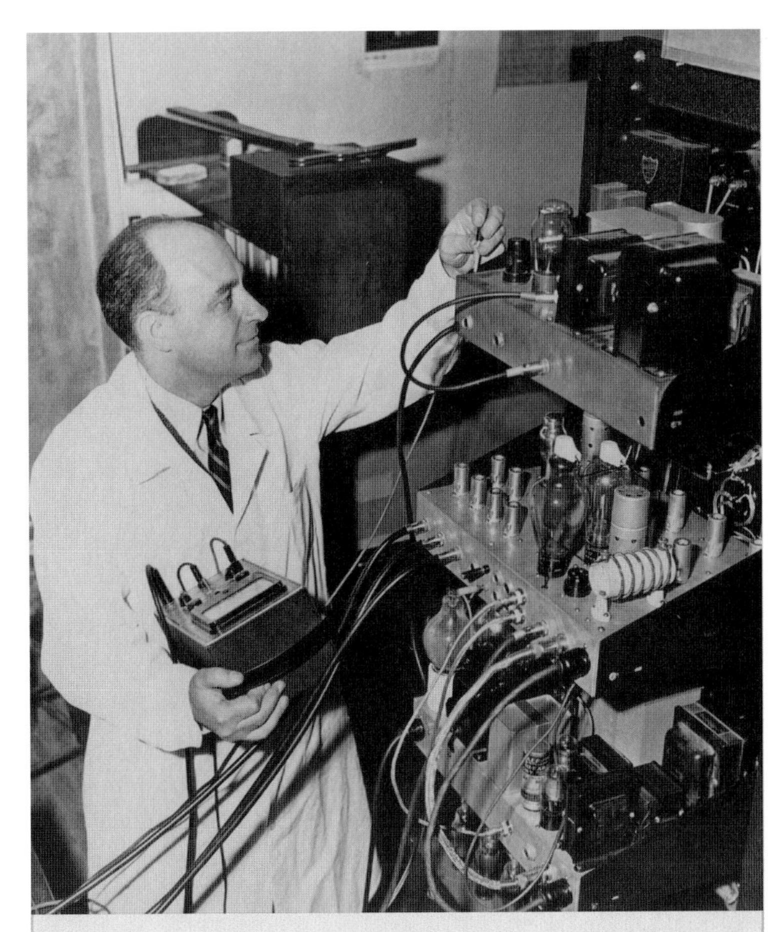

The Italian-American physicist Enrico Fermi. He built the first atomic reactor in 1942 at the University of Chicago.

BOMBARDING THE NUCLEUS WITH NEUTRONS

In his work with nuclear scattering experiments, the Italian physicist Enrico Fermi (1901–54) found that neutrons were more effective in initiating nuclear reactions (reactions involving particles and nuclei in which energy is released or absorbed) if they had little speed, or energy, when they encountered the nucleus. With a low energy they stayed in the vicinity of the nucleus longer and reacted with it better.

Fermi began a program of bombarding nuclei with neutrons to see what type of nuclear reactions would be initiated. He was particularly interested in the heaviest known nuclei (at the time), uranium. He thought that if he bombarded it with neutrons he might be able to create a new nucleus, or element, beyond uranium. And for a while he was confident that he had, and he even named it uranium-X. What Fermi had inadvertently caused was a new phenomenon called fission, which would soon play a large role in physics. Unfortunately, he did not check his experimental products closely enough and failed to realize what he had done.

Fission

In 1939 Hahn and Strassmann of the Kaiser Wilhelm Institute in Berlin bombarded uranium with neutrons and were surprised when they discovered barium in the products of the bombardment. Barium is only half as massive as uranium, and it seemed unlikely that it could have been created in any of the reactions. Hahn wrote to physicist Lise Meitner (1879–1968) to see what she thought of the unusual result.

Meitner shared the letter with her nephew Otto Frisch (1904–79) who was visiting (he was also a physicist). They tried to imagine what would happen when a uranium nucleus absorbed a neutron, and decided that it might become unstable and begin to oscillate, and if so it might break in half. The resulting nuclei would be only half as massive as uranium, and this could account for the barium. Meitner made some calculations, and it was soon clear that this is indeed what had happened. They called the process nuclear *fission*, which means "breaking apart."

Meisner and Frisch passed the information on to Niels Bohr, who was just leaving for the United States. The discovery was formally announced soon after Bohr reached New York, and the news spread rapidly.

The Hungarian physicist Leo Szilard (1989–64), who had just fled the Nazi regime in Germany, heard about it and worried that the Germans would use it to build an atomic bomb. He wanted to make sure that the Americans built one first, so he talked Einstein into signing a letter he had written to President Roosevelt. The letter was delivered to Roosevelt, and a year or so later the Manhattan Project was initiated, with Fermi as director. In December 1942 a "slowed down" version of an atomic

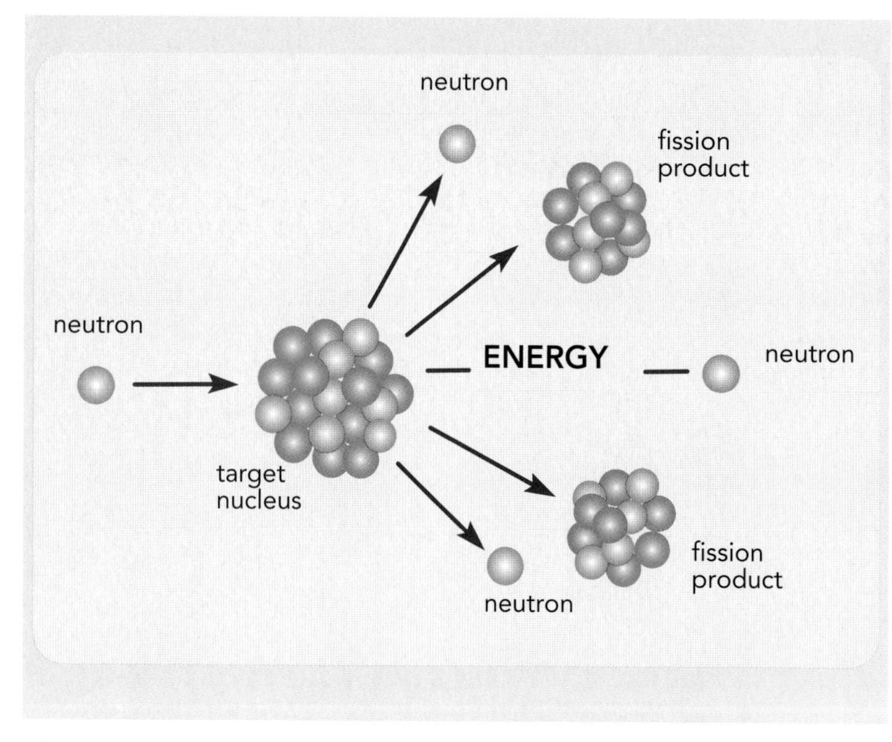

Above: A simple representation of fission. A neutron hits a target nucleus causing it to fission. As it breaks apart it releases other neutrons that fission other nuclei. Top left: The stacks of a nuclear power plant.

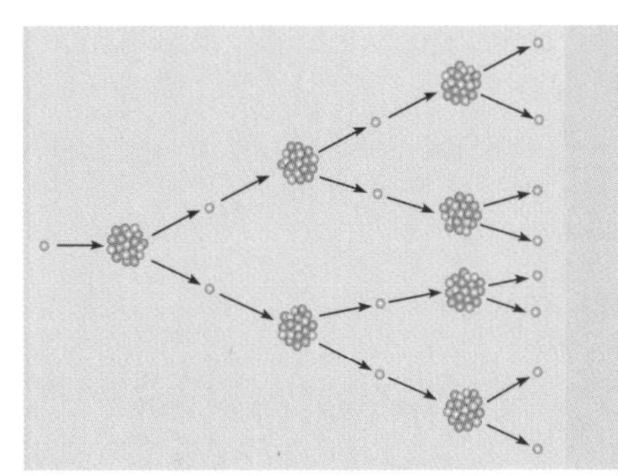

A representation of a chain reaction caused by nuclear fission.

bomb was built and tested at the University of Chicago. Los Alamos, New Mexico, was then selected as the site for the design and assembling of the first bomb. It was tested in 1945 near Alamogordo, New Mexico, and within weeks was used on Hiroshima and Nagasaki.

BUILDING THE BOMB

Natural uranium has two isotopes, U-238 and U-235, and only U-235 is fissionable, so the first step in building the bomb was to separate out the U-235. Only 0.7 percent of natural uranium is U-235, and therefore a lot of uranium would be needed. Furthermore, the two isotopes are very similar and could not be separated chemically; physical processes such as diffusion and centrifugation had to be used.

The second step was determining how much U-235 would be needed. The answer depended on understanding chain reactions. To see what a chain reaction is, assume we have an assemblage of U-235 nuclei. If a neutron is projected at it, one of the nuclei will fission, giving off two or more neutrons in the process, and each of them will cause other nuclei to fission. This will proceed like this: $1 \rightarrow 2 \rightarrow 4 \rightarrow 8 \rightarrow 16 \rightarrow 32 \rightarrow 64$ and so on, quickly reaching into the billions. Amazingly, it takes less than a millionth of a second for this to happen.

One of the major problems, however, is keeping the reaction going, and this is where the size, or amount of U-235 needed, comes in. If the size of the uranium is too small (we can assume it is a sphere), many of the neutrons will escape through the surface without causing other nuclei to fission. If it is too large, on the other hand, it will be blown apart quickly and most nuclei will not fission. The amount that is required is called critical size.

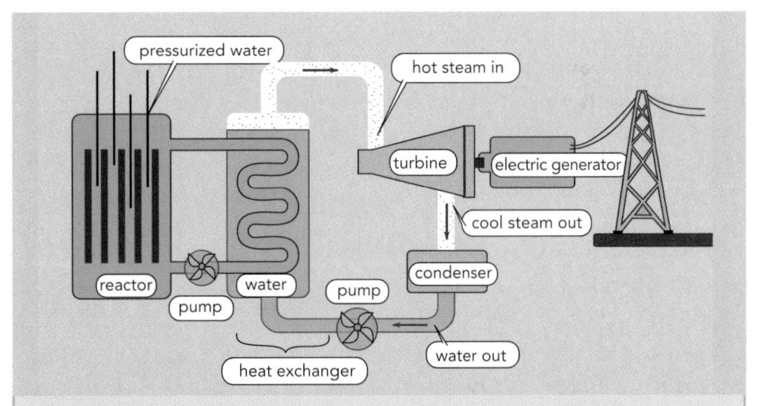

This diagram of a nuclear reactor shows how it generates electric energy using steam.

NUCLEAR REACTORS

A nuclear reactor is a "slowed down" version of the atom bomb. In this case, the energy does not come all at once but is spread out over time and is therefore controllable, and usable. Reactors require an external source of neutrons, so it is not possible for reactors to undergo a nuclear explosion. For a reactor we need what is called a *moderator*—a material that slows down neutrons. Two of the best moderators are graphite and heavy water. In a reactor, the uranium (or other fissionable material) is usually in the form of rods, which are placed in the moderator and interspersed with other rods called control rods (they are usually cadmium). The nuclear reactions within the reactor can be sped up or slowed down by adjusting the position of the control rods. A coolant such as water is also usually used to keep the reactor cool.

Nuclear reactors are used extensively for power, but they are also used for other purposes. Among other things, reactors provide a source of neutrons, induce various nuclear reactions and make fissionable material, and prepare radioisotopes for use in medicine.

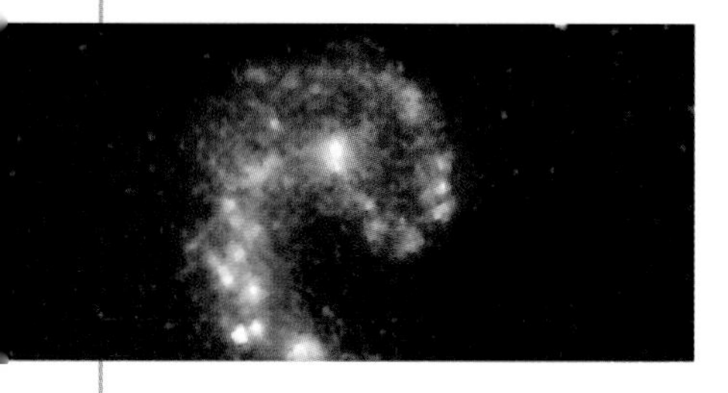

Fusion

The fissioning of nuclei is not the only way that nuclear energy can be generated. A process in which light nuclei come together and fuse also produces a tremendous amount of energy. This is, in fact, the process that occurs in stars (including our Sun), and it is also the process that is used in making hydrogen bombs. In the latter case the very lightest element, hydrogen, and its two isotopes, deuterium and tritium, are involved. In our Sun, four hydrogen nuclei come together and fuse to form a helium nucleus, and in the process energy is radiated off into space. Earth receives some of this energy in the form of light and heat. But the reaction that occurs in the Sun is exceedingly slow. For a hydrogen bomb, something much faster is needed, and as it turns out, several reactions involving deuterium and tritium are much faster.

HYDROGEN BOMBS

All that is required to make a fusion reaction is extreme heat, and the best way to get it is to explode an atomic bomb. A hydrogen bomb therefore consists of a large chamber of liquid hydrogen (or deuterium) and an atomic bomb. When the atomic bomb explodes it causes the hydrogen to fuse, which creates a much larger explosion. In fact, there is almost no limit to the explosive power of a hydrogen bomb. As we saw earlier, atomic bombs are restricted to a certain size because of the critical size of the mass of uranium. Hydrogen bombs, on the other hand, have no restriction, so very powerful bombs can be built. As a comparison, the atomic bomb dropped on Hiroshima was a 10 kiloton bomb (equivalent to 10,000 tons of TNT), but hydrogen bombs of 20 megatons (20 million tons) and even 50 megatons have been built.

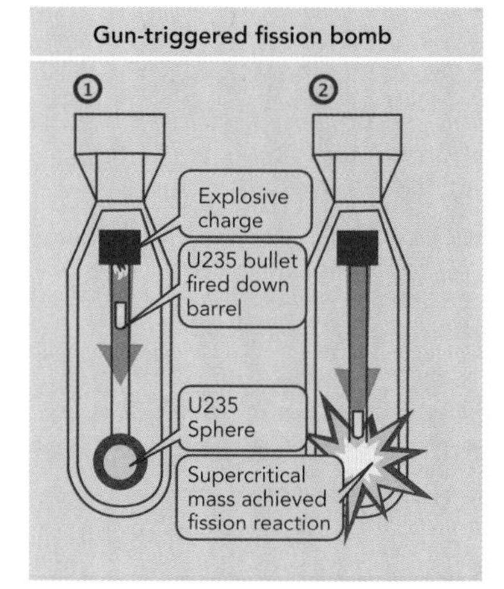

Above: An atomic bomb similar to the one dropped on Hiroshima in 1945. In figure 1, two subcritical pieces of U235 are kept separated. In figure 2, an explosive charge brings them together to cause a chain reaction and explosion. Top left: A group of stars being born.

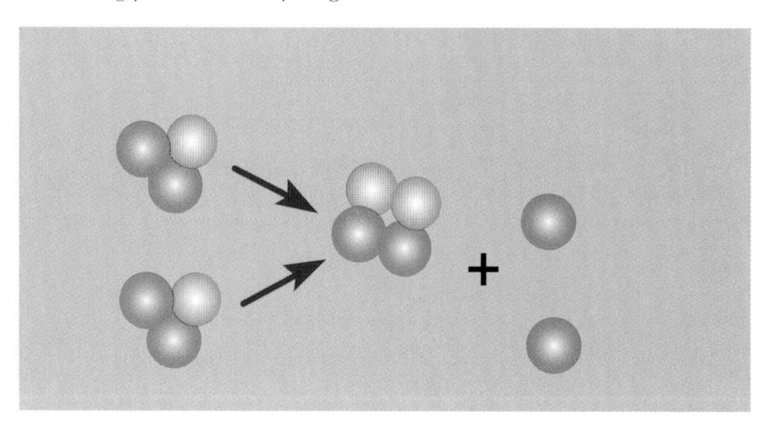

A simplified version of the proton-proton reaction that takes place in the Sun. Two helium-3 nuclei collide to form helium-4 with the release of two protons. The first steps of the process are not shown.

FUSION IN STARS

In our Sun, energy is produced when four hydrogen nuclei fuse into a helium nucleus. The process, however, is not merely that of four hydrogen nuclei coming together; it is much more complicated. The cycle for the Sun, and many other stars, is known as the *proton-proton cycle*, and it was discovered in the 1930s by the physicist Hans Bethe (1906–2005). For this cycle to operate, the core temperature of the star has to be about 10 to 15 million degrees Kelvin. It begins with the creation of a deuteron (a deuterium nucleus). When this deuteron is hit by a proton, it becomes helium-three (^{3}He, an isotope of helium). Finally, two ^{3}He nuclei fuse to form ^{4}He. The process takes millions of years, but there are billions upon billions of particles going through the cycle at any time, so a large amount of energy is produced continuously. In fact, 512 million metric tons of hydrogen are converted to 508 million metric tons of helium every second in the Sun. The 4 million metric tons that is missing is the mass that is converted to energy (according to Einstein's formula).

The proton-proton cycle is not the only one operating in stars. If the core temperature of the star is greater than 15 million degrees, a cycle called the *carbon cycle* operates. Hydrogen is also converted to helium in this case, but carbon acts as a kind of catalyst. In extremely hot stars another cycle, called the *triple alpha cycle*, operates. In this case, three helium nuclei (alpha particles) come together to form a carbon nucleus.

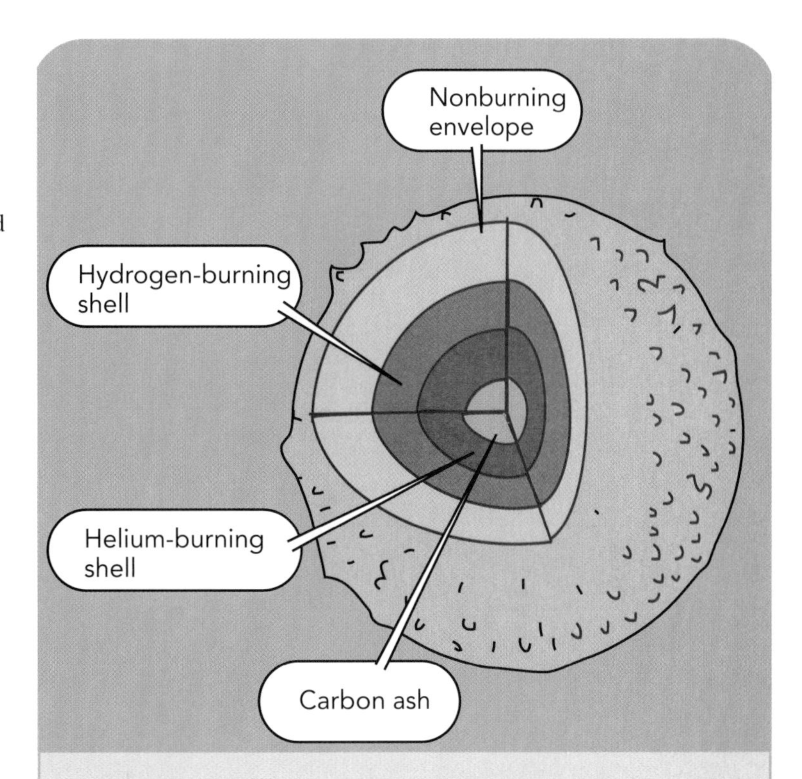

A cross-section of the Sun showing the "burning shells" near its core. The outermost region is the nonburning envelope of hydrogen. The first shell is the hydrogen-burning shell, beneath it is a helium-burning shell, and at the center is carbon ash.

CREATING NEW ELEMENTS AND NEW STARS

Where did the elements in the universe come from? For years, this was one of the major problems in astronomy. It was shown early on that most elements could not have been produced when the universe was formed. The problem was solved when astronomers began theorizing about what happens in the core of very massive stars. Hydrogen "burning" (as it is usually called) leaves helium as ash, and as the temperature of the core increases, other, hotter-burning cycles are initiated. The ash that is created in each case falls to the core of the star, but because of the increased pressure it encounters there, it soon begins to burn, until finally the star is burning on many levels. If you could see the core of a very massive star, you would see successive shells of burning hydrogen, helium, carbon, neon, oxygen, silicon, and iron (see diagram). When the star finally has an iron core, however, it can go no further. At this stage the star becomes unstable and explodes as a *supernova*, and in the process, all of the elements beyond iron, out to the heaviest elements in the universe, are generated. The gaseous remnants of supernovas eventually condense and form new stars, and these new stars are enriched with heavy elements.

CHAPTER 12

THE STANDARD MODEL

Left: Particle interactions as seen in a modern detector. Top: The Cosmotron accelerator at Brookhaven National Laboratory. In the early 1950s it was the most powerful accelerator in the world. It was the first synchrotron accelerator to provide an external beam of particles for experimentation outside the accelerator itself. Bottom: One of the control rooms at Fermilab's Tevatron particle collider.

An understanding of the interaction of particles and fields (such as the electromagnetic field) was an important objective for physicists in the years after the discovery of quantum mechanics. But many of the particles of nature were traveling near the speed of light when they interacted, so they had to be treated relativistically. Schrödinger's equation, however, was not relativistic, so physicists began looking for a relativistic extension of it. Paul Dirac (1902–84) of England made the breakthrough in 1928 with an equation that is now referred to as Dirac's equation. It applied only to particles of spin with a value of one-half, which included the electron—so the equation could describe interactions involving relativistic electrons.

Several breakthroughs came in the years that followed, which led to the development in the 1970s of what is known as the standard model. This is the current theory that describes almost all the known processes in particle physics.

Before the Standard

Despite its successes, Dirac's equation gave a strange result for electron energies. It predicted that in addition to positive energies, the electron could also have negative energies. This meant that transitions between the positive and negative energy states were possible, yet they had not been observed. In 1929 Dirac postulated that the "sea" of negative energy states was filled, and therefore transitions from positive to negative states were impossible. But the reverse—transitions from negative states to positive states—could still occur, which would leave an observable "hole" in the negative sea. It would look exactly like an electron, but would have a positive charge. This meant that a positive electron would exist. A few years later just such a positive electron, or *positron*, was discovered by Carl Anderson of the California Institute of Technology.

The positron is now considered the *antiparticle* of the electron, and if it collides with an electron the particles annihilate each other with the release of energy in the form of photons. This discovery has led scientists to find that all particles have antimatter partners. Corresponding to the proton, for example, there is an antiproton.

FEYNMAN AND QUANTUM ELECTRODYNAMICS

Using Dirac's theory, physicists could now calculate how electrons, photons, and other particles would interact, and what the results would be. The simplest, or first-order, calculations were in good agreement with experimental results, but when higher-order, or more accurate, calculations were made scientists expected a small correction to the first-order results (making them more accurate). But, to their surprise, the second-order calculations gave an infinite result. This meant that highly accurate calculations of the interactions between particles were not possible.

In 1948, three physicists, working independently, found a way to make accurate calculations. They were Julian Schwinger (1918–94) of Harvard University, Richard Feynman (1918–88) of the California Institute of Technology, and Shinichiro Tomonaga (1906–79) of Japan. The method they used is referred to as *renormalization*, and the theory they developed is called *quantum electrodynamics*, or QED. Each physicist's method was slightly different, but it was Feynman's method that eventually became the most widely accepted. He used diagrams of the interactions to assist in writing down the mathematical formulas he needed. For example, if two electrons interacted, the diagram would appear as below.

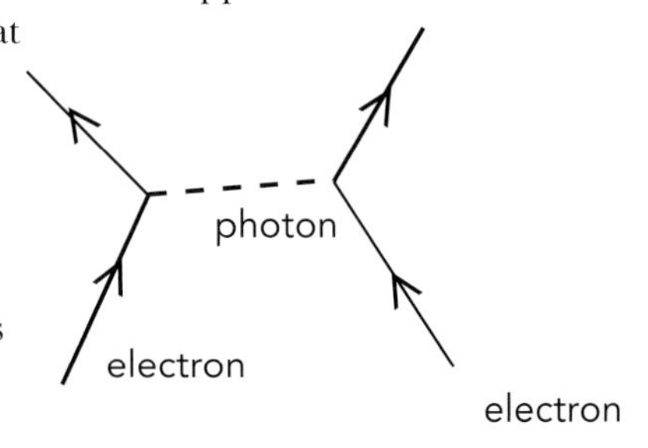

Above: A Feynman diagram showing the collision of two electrons. Top left: Paul Dirac, who first explained electron spin with his theory of the electron. The same theory predicted the existence of antiparticles such as the positron, the antiparticle to the electron.

A closer look at this diagram shows that it violates conservation. Initially there are two electrons present, and then suddenly there are two electrons and a photon. Where does the energy for the creation of a photon come from? The answer to this question was an important step in the technique developed by the three men, and it came via the uncertainty principle. According to this principle, we cannot measure two variables such as position and momentum *exactly* (at the same time). In other words, there is a "fuzziness" associated with particles on this level . Strangely, this fuzziness is to our advantage, for it means that energy can be "borrowed" under its veil, as long as it is paid back very quickly.

Consider an electron-electron scattering, with the exchange of photons. We assume that as two electrons approach each other, each has a cloud of what we call "virtual" photons associated with it. When the electrons get close to each other, photons from the two clouds begin moving back and forth, and the closer the electrons get, the larger the number of photons. As the electrons pass and the distance between them increases, the number of exchanges decreases. In a Feynman diagram we show this interaction merely as a dotted (or wavy) line (above left).

There are, of course, many different reactions that can be represented by Feynman diagrams. Another is the absorption of a photon by an electron, and its emission after a very brief interval. It is referred to as Compton scattering. Both it and the electron-electron scattering are first-order diagrams that allow first-order calculations. But it is the second-order calculations that present a problem. Feynman showed that second-order diagrams can help to clarify these calculations. These diagrams are more complicated, as seen in the example at right.

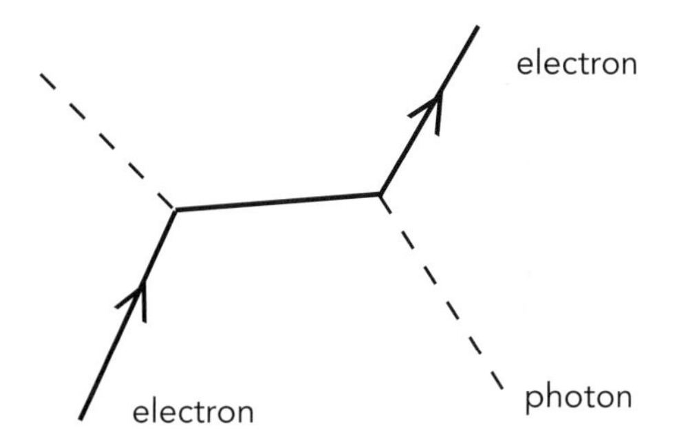

Above: This Feynman diagram shows the absorption of a photon by an electron and its emission a short time later. Below: A second-order Feynman diagram.

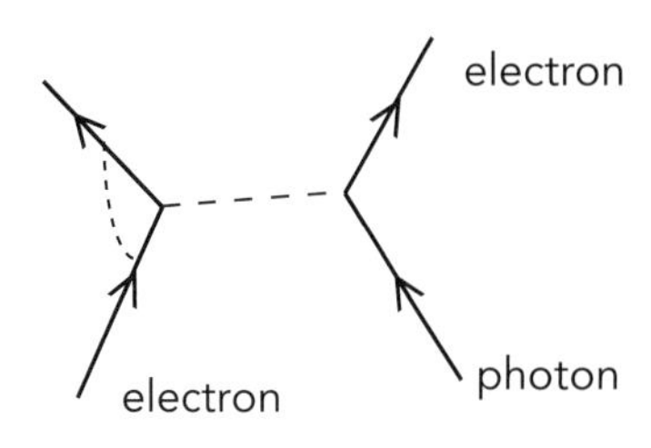

FEYNMAN, THE MAN

Born in New York City in 1918, Richard Feynman received his bachelor's degree from Massachusetts Institute of Technology in 1939 and his PhD from Princeton in 1942. A genius of the first order, Feynman amazed people with the complex calculations he could do in his head. He liked to bet people that he could solve any problem within 60 seconds that they could state in 10 seconds. And he usually won. He was also well known for his high jinks and antics. He loved to play the bongo drums and frequently played them at three o'clock in the morning, much to the dismay of his neighbors.

During World War II, Feynman worked at Los Alamos, and during that time (in addition to making important contributions to the bomb project) he developed considerable skill in cracking safes. To the dismay of many ranking officials, he would "borrow" secret documents from a safe and leave a note saying that he had left them in the top drawer of a nearby desk. Unlike most scientists, Feynman did not require a quiet place to do his work; in fact, he was frequently seen in topless bars working away on his equations, with music blaring in his ear.

Richard Feynman, the American physicist that developed the theory of quantum electrodynamics (QED). He also worked on the atomic bomb at Los Alamos.

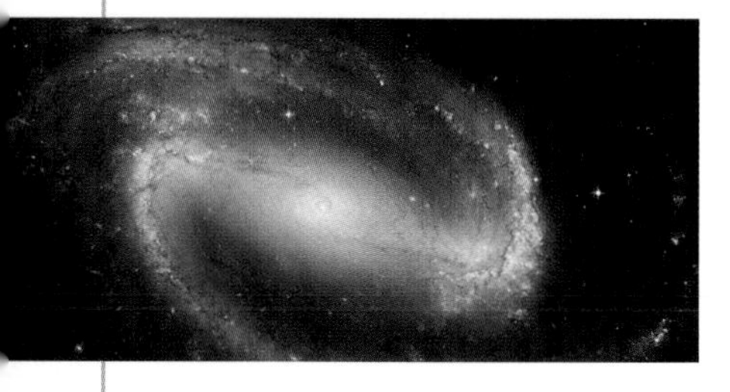

Fundamental Forces

Four fundamental forces (or fields) exist in nature. We have already met three of them—the electromagnetic force, the gravitational force, and the strong nuclear force. The fourth force is called the *weak nuclear force*, and it is mainly involved in radioactive decay. The relative strengths of these forces are as follows:

Gravitational force	1
Weak nuclear force	10^{25}
Electromagnetic force	10^{37}
Strong nuclear force	10^{39}

Gravity is, of course, the most familiar of these forces. And to us it appears to be very strong. After all, it holds us on the surface of the Earth, and it keeps the planets in orbit around the Sun. Strangely, though, it is by far the weakest of the four forces. The reason it seems so strong is that it is usually associated with large bodies, or large masses. If you tried to measure the gravitational force between small things, however, such as two elementary particles, you would find it to be exceedingly weak. It is also important to note that the gravitational force is a long-range force in that it acts on bodies that are far apart, as well as on bodies that are close together.

The second most familiar force is the

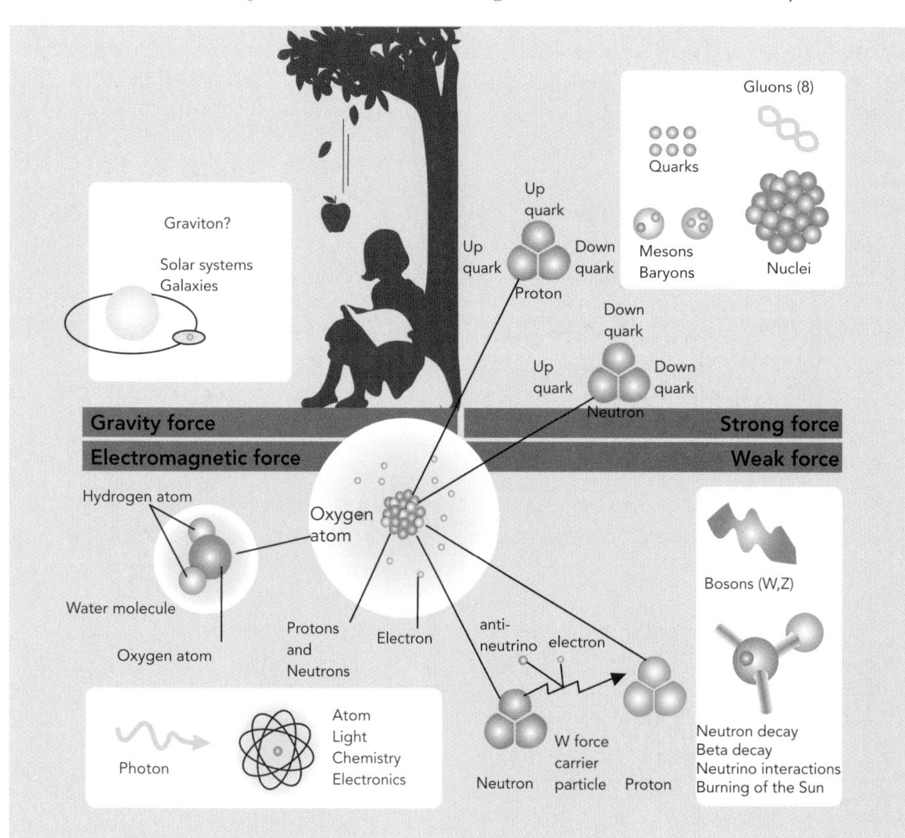

Above: The four fundamental forces of nature (clockwise from upper left): gravity; the strong nuclear force that holds nucleons in the nucleus; the electromagnetic force that holds atoms together; and the weak nuclear force that plays an important role in radioactive decay. Top left: A barred spiral galaxy. This image was taken by the Hubble telescope. The form of this galaxy is a result of gravitational forces.

electromagnetic force. It holds atoms together. It is also long-range, like the gravitational force, but it differs in an important respect: The gravitational field is always attractive, but the electromagnetic field can be both attractive and repulsive (unlike charges attract, similar ones repel).

NUCLEAR FORCES

The other two forces, the strong and weak nuclear forces, are quite different from the two above forces in that both of them are short-range. They are strong over a very short distance, about the size of the nucleus, but reduce to zero beyond that. The weak nuclear force is seen in what is called beta decay, which is the conversion of a neutron to a proton, with the emission of an electron. An exchange particle, called the W particle, is involved in the process. More exactly, beta decay consists of a neutron emitting a W particle and becoming a proton, with the W particle (which has an extremely short lifetime) being converted almost immediately to an electron, with the emission of an antineutrino.

Each of the four forces has an exchange particle. They are:

Gravitational field	graviton
Weak nuclear field	W particle
Electromagnetic field	photon
Strong nuclear field	meson

QUANTUM GRAVITY AND GRAVITATIONAL WAVES

Quantum theories have been developed for three of the fundamental forces, and in each case the exchange particle is well known and has been observed experimentally. So far, however, we do not have a quantum theory of gravity, which would be needed to mathematically describe the interactions of masses via gravitons. The theory that describes gravity, namely general relativity, is a geometric theory. Many attempts have been made to "quantize" it, but none has been successful. This means that we do not have the proper tools for dealing with gravitons. Nevertheless, we can picture what they would be like and how they would act. The graviton would no doubt be similar to the other exchange particles but would act between matter, so that when two masses approached each other, gravitons would pass back and forth between them. Like the electromagnetic field, gravity is long-range, so when the masses are close, many gravitons would be exchanged; but as the distance increased, their number would decrease.

When an electric charge oscillates an electromagnetic wave is emitted. When a mass oscillates a gravitational wave should be given off. Since gravity, however, is much weaker than the electromagnetic field, gravitational waves would be exceedingly weak, which is why they have not yet been detected. But instruments that will be able to detect these waves are currently under construction, one of which will involve components that will orbit the Sun ahead and behind the Earth.

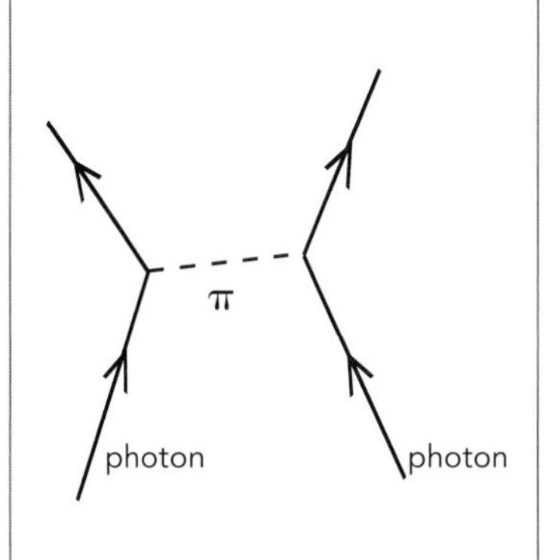

Feynman diagrams were not only used in quantum electrodynamics. For years they were also important in the strong interactions involving protons, mesons and so on. The above reaction is the collision of two protons with the exchange of a π^0 meson.

FEYNMAN DIAGRAMS FOR OTHER FORCES

We saw that Feynman developed a very useful diagrammatic technique for dealing with the electromagnetic interactions. But it can also be used for the other three interactions. For example, when two protons interact, we have the exchange of π mesons, which can be represented as shown above.

This means that we can visualize protons, neutrons, and so on as also surrounded by virtual particles; in this case it is pions. It is important to point out, however, that recent developments have given us a different view of the strong interactions. In the same way we can draw diagrams for the exchange of W particles, and even gravitons—although gravitons have never been detected experimentally.

Fundamental Particles

Soon after the positron was discovered, scientists realized that all particles had antiparticle partners. And in the years that followed the discovery of the meson, more and more particles were discovered, until finally the number approached 300. Physicists began to wonder if all these particles were truly fundamental. It was possible that some of them might be made up of other, more fundamental particles.

The first step toward determining which were truly fundamental was to divide the particles into two groups: the *hadrons* (heavy particles) and *leptons* (light particles). The hadrons were, by far, the largest group, which included the proton and the neutron. The leptons were a relatively small group, consisting of the electron and positron, the muon and its antiparticle, and a particle called tau and its antiparticle. Associated with each was a neutrino, which is a particle with no charge and little or no mass.

In the early 1960s two physicists from the California Institute of Technology, Murray Gell-Mann (b. 1929) and George Zweig (b. 1937), independently suggested that the hadrons were made up of more fundamental particles. Gell-Mann called them *quarks*, and Zweig called them aces, but soon everyone was referring to them as quarks. According to

Above: An aerial view of the Fermilab accelerator in 1999. The main injector is in the foreground, the Tevatron is in the background. Top left: An artist's conception of an electron and positron colliding at high speed. This is a good example of matter annihilating antimatter with the release of pure energy in the form of photons.

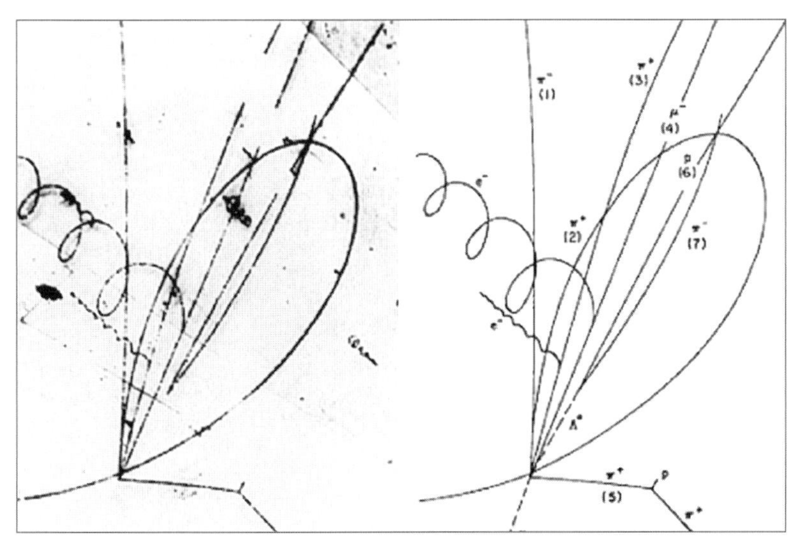

Photograph (left) and diagram based on it of a particle event in a cloud chamber at Brookhaven National Laboratory. This photograph led to the discovery of the charmed quark.

Gell-Mann there were three of these quarks, which he called up (u), down (d), and strange (s), along with their antiparticles ū, đ, and š. (We now refer to each of the types as *flavors.*) They did not have a unit charge, and their charges differed: The up quark had a charge of +2/3 (times the electronic charge); the down and strange quarks had a charge of -1/3.

Gell-Mann postulated that baryons were composed of three quarks, and mesons were composed of a quark and an antiquark. Both baryons and mesons are hadrons. Furthermore, the quarks were confined to "bags" and never escaped from them. But the question of what held the quarks together remained.

THE COLOR FORCE

To overcome the above problems, O. W. Greenberg of the University of Maryland suggested in 1964 that the three flavors of quarks each came in three "colors" (no relationship to the usual meaning of color). And of particular importance, the three colors had to add up to white. They are usually designated as: red, green, and blue. Since color was actually a force, an exchange particle was needed; it was called the *gluon* (a pun—it's the "glue" that holds the quarks together). Each of the gluons has two colors associated with it.

In 1974 a new particle called ψ/J was discovered, and it had a property called *charm.* This meant that there was a fourth quark called charm (c). Since then, two more quarks have been predicted to exist and subsequently found—the bottom (b) and top (t). In all, then, there are now six quarks along with their antiparticles, and each can have one of three colors.

Quarks have never been observed. They are trapped inside the "bags" that hold them, and according to quark theory we will never see them. Due to developments in particle theory, we now view the strong interactions differently. Where it was once assumed they were due to the exchange of mesons, we now know that mesons are composed of quarks, and quarks are held together by gluons. Therefore, it is gluons that are passed back and forth when two nucleons are close to each other, only giving the appearance of an exchange of mesons.

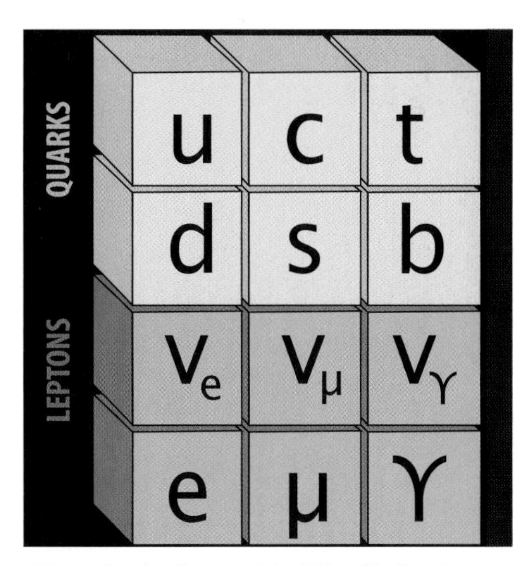

The twelve fundamental building blocks of matter. The top six are the up, down, charmed, strange, top, and bottom quarks. The bottom six are the electron, the muon, the tau, and their neutrinos.

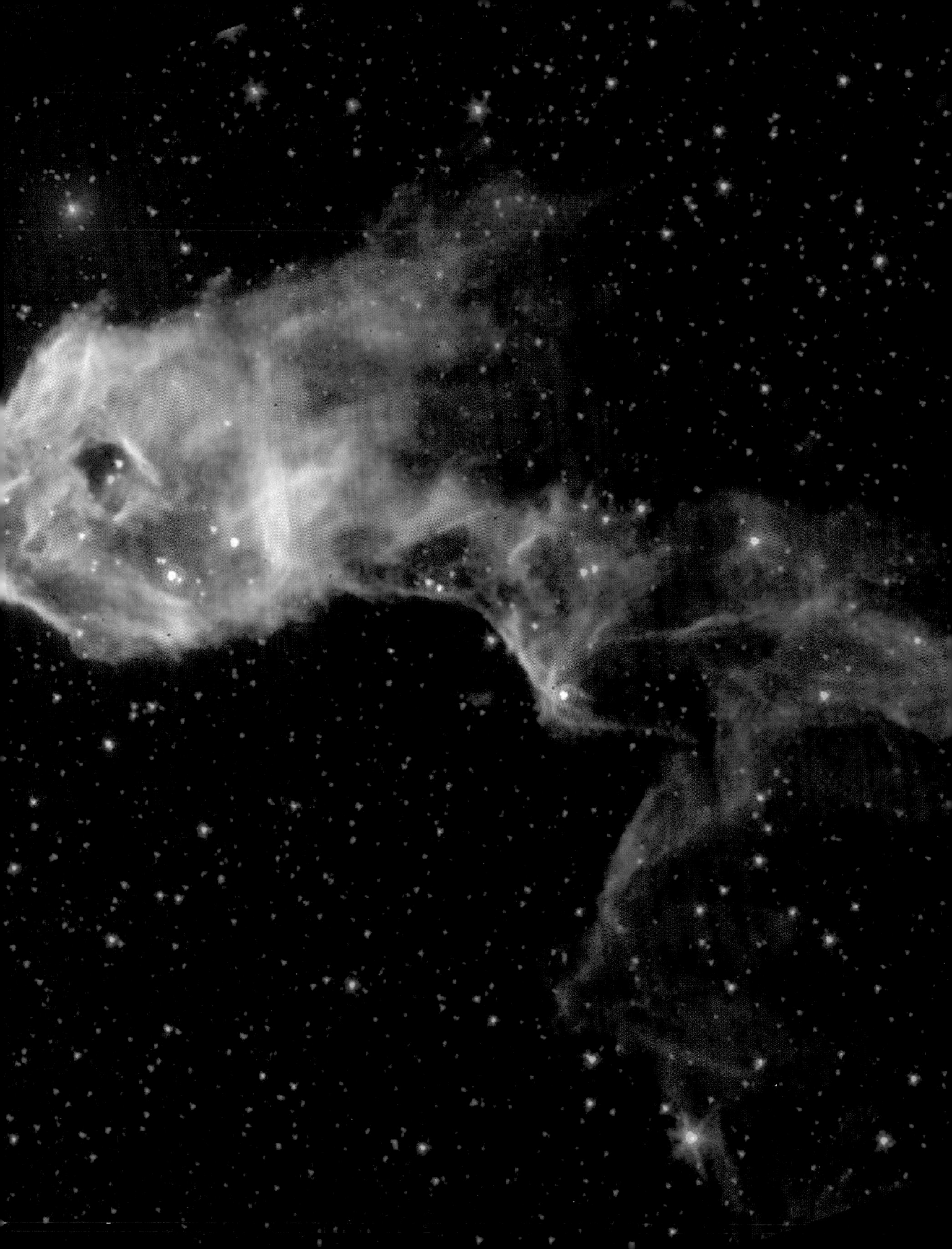

CHAPTER 13

MOGULS OF MODERN COSMOLOGY

Left: A large nebula known as the Elephant's Trunk nebula, in which many new stars are forming. Top: Artist's conception of a supermassive black hole at the center of a galaxy. Bottom: The edge-on view of a galaxy. Considerable dust can be seen in the spiral arms.

Cosmology is the study of the overall structure of the universe and is also concerned with the age and size of the universe, how it began, and how it has evolved. Technically it is a branch of astronomy, but many of the major advances in cosmology have been made by physicists. Newton can be thought of as the first cosmologist, as he used his theory of gravity to formulate a model of the universe. Einstein was also a cosmologist in that he devised one of the first successful cosmologies.

Many other people have contributed to cosmology. Georges Lemâitre (1896–1966) of Belgium gave us the first theory that explained the big bang. Lorentzian de Sitter (1872–34) of the Netherlands introduced an interesting model shortly after Einstein published his model. George Gamow (1904–68) of the United States also made important contributions to cosmology—particularly in relation to our understanding of the early universe. And finally, Fred Hoyle (1915–2001) of Cambridge University gave us an important alternative to the big bang theory called the steady state model. In the twentieth century, great leaps were made toward a better understanding of the universe through the work of several notable physicists. Some of them continue this work today.

Hubble Measures the Galaxies

Born in Marshfield, Missouri, Edwin Hubble (1889–1953) was an outstanding scholar through high school and university who also excelled in many sports, including track, basketball, boxing, and rowing. He studied physics at the University of Chicago, but when he won a Rhodes scholarship to Oxford University he switched to law. Upon graduation in 1913 he returned to the United States and set up a law practice in Louisville, Kentucky. He soon became dissatisfied, however, and finally decided to return to the University of Chicago and major in astronomy. While there he studied the so-called white nebulae using the large telescopes of Yerkes Observatory. At the time no one was sure what these nebulae were. Some appeared to be huge gaseous clouds, but there was considerable speculation that some were composed of stars that could not be seen because they were so far away.

When Hubble graduated he was offered a job at Mount Wilson Observatory in California, but the United States had just entered World War I, and he decided to enlist. After the war he went directly to the Mount Wilson Observatory and began one of the most extensive studies of the white nebulae that had ever been done.

Edwin Hubble, discoverer of the expansion of the universe.

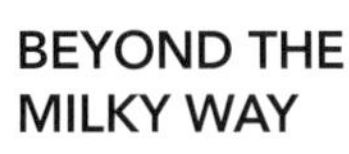

BEYOND THE MILKY WAY

A few years earlier Vesto Slipher (1875–1969) of Lowell Observatory in Arizona had studied the spectra of a large number of white nebulae and noticed that many of them were redshifted. Redshift is the shift in spectral lines toward the red end of the spectrum that is observed when a galaxy is moving away from us. However, Slipher noticed that a few were blueshifted, which indicated they were approaching us. Slipher did not arrive at any conclusion as a result of his study, and by 1920 he had reached the limit of his relatively small telescope and went on to other projects.

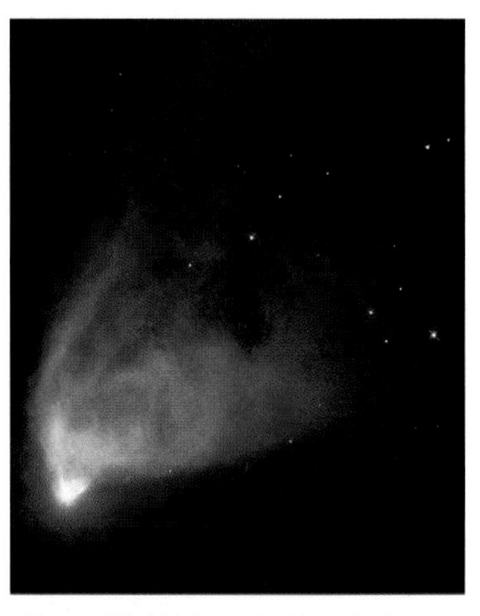

Above: Hubble's variable nebula. It is of particular interest because it varies in brightness over time. Top left: The Andromeda galaxy. This is one of the galaxies Hubble was studying when he discovered that galaxies were "island universes" of stars separate from the Milky Way.

Hubble had access to the largest telescope in the world, namely the hundred-inch Hooker telescope at Mount Wilson, and he took advantage of it. His work at Yerkes had convinced him that many, if not most, of the white nebulae were "island universes of stars," and he set out to prove it. He took long exposures of some of the larger nebulae in the sky, including one in the constellation of Andromeda, called the Andromeda nebula. The Andromeda nebula was a spiral with long, distinct arms. Hubble concentrated on the arms and was eventually able to distinguish individual stars in them.

A large nebula, or gas cloud, that lies within the galaxy called M33. Hubble observed this galaxy in 1924.

VARIABLE STARS

Hubble then set about trying to determine the distance to these stars, and as it turned out, he was very lucky. Some of the stars varied in brightness in a regular way; in other words, they were variable stars—a particular type of variable star called *Cepheids.* Cepheids had been studied for many years, and it was known that they satisfied a period-luminosity relationship. This relationship was discovered by Henrietta Leavitt (1868–1921) and calibrated by Harlow Shapley (1885–1972), both of Harvard University. It allowed astronomers to determine the distance to a Cepheid by measuring its period. Using the relationship, Hubble determined the distance to the Andromeda nebula to be slightly over 900,000 light years. This was considerably larger than our galaxy (the Milky Way), and it meant that the Andromeda nebula was well outside of it; in other words, it was a separate system.

By late 1924 Hubble had detected and measured 12 Cepheid variables in the Andromeda galaxy and another 22 in a nearby galaxy called M33 (from a table called Messier objects). In both cases the distance was slightly over 900,000 light years (this was later corrected to 2 million light years). Hubble measured the distance to several other galaxies, and it was soon clear that the universe was made up of galaxies—huge systems of stars, like our Milky Way galaxy.

The Expanding Universe

Using Cepheids, Hubble was able to determine the distance to six nearby galaxies. But in more distant galaxies, Cepheids were too small to be visible. Some stars, however—the largest in the galaxy—were still visible. Hubble made the assumption that the brightest stars in all galaxies were about the same brightness. Using this in conjunction with the well-known law that tells us how fast the brightness of a light source drops off with distance (an inverse-square law), he was able to obtain the distance to 14 more galaxies. This gave him fairly reliable distances to 20 galaxies. The redshifts of most of these galaxies had been measured by Slipher, and Hubble used them to plot redshift (or equivalently, recessional velocity) versus distance for the galaxies. Hubble's observations left no doubt: There was a linear relationship, which meant that the farther away the galaxy, the faster it was receding from us. Hubble published his results in 1929. At this point there were still uncertainties in his results, but he was determined to press on by measuring galaxies even farther out.

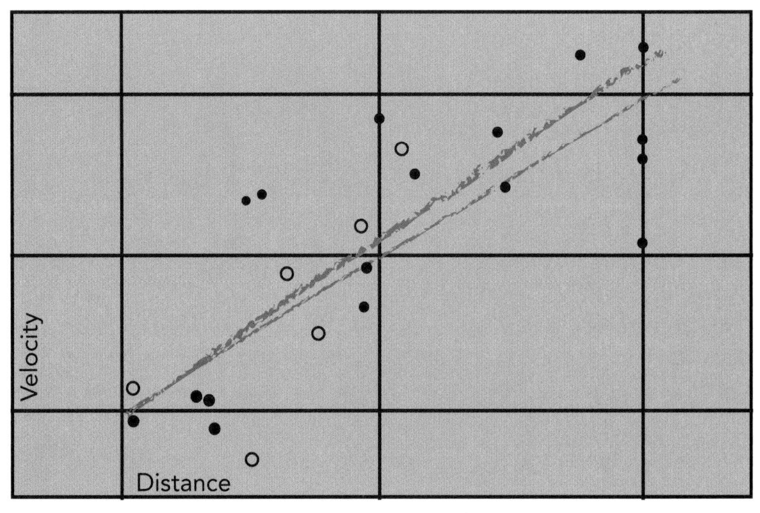

Above: A Hubble plot of distance versus velocity for galaxies. Note that the relationship is linear, but there is considerable scatter of the points due to various uncertainties in measurement. Top left: A cluster of galaxies. The large elliptical galaxy at the center of the photo is surrounded by numerous smaller spiral galaxies.

Hubble at the eyepiece of a new Schmidt telescope at Palomar Observatory in California, 1949. Hubble had used another telescope at the observatory to determine the distances and velocities of neighboring galaxies.

THE COSMIC LADDER

He now had to begin measuring redshifts, and most of this work was done by his assistant, Milton Humason (1891–1972). Hubble spent his time determining distances. As we saw, after he ran out of Cepheids, he used the brightest stars in galaxies; but eventually he got to the point where stars were no longer visible. He then turned to the brightness of the galaxy itself; in particular, he concentrated on clusters of galaxies and made the assumption that the largest in each case were of similar size and absolute brightness. In this way he constructed a "cosmic ladder" to the outer reaches of the universe.

By 1931 Humason had measured the redshifts of 37 more galaxies, and Hubble had determined their distances. They were added to the original plot of redshift versus distance. In his 1929 paper Hubble had measured and plotted galaxies out to about 6 million light years. By 1931 he had increased this to 100 million light years—16 times farther out. He published his second paper in 1931. There appeared to be no doubt now: The universe was indeed expanding. The farther a galaxy was from us, the faster it was receding from us.

THE HUBBLE CONSTANT

In 1936 Hubble published *The Realm of the Nebulae*, in which he presented his results. He had now explored galaxies out to 240 million light years, and it was clear that recessional velocities were increasing linearly. The slope of the plot in the graph is now known as the Hubble constant, and it is usually represented by H. Its reciprocal ($1/H$) gives us an estimate of the age of the universe, but it is only an approximate estimate in that it does not take into consideration any accelerations or decelerations that may have occurred in the past.

Looking at the plot, it is easy to see that if you go out far enough, the recessional velocity will eventually reach the velocity of light (a galaxy twice as far out is receding twice as fast). But according to Einstein's theory of relativity, the velocity of light is the limiting velocity in the universe. This implies that galaxies beyond this point can't exist, and therefore this is the end of our observable universe. In the years that followed Hubble's original discovery, several corrections and adjustments were made, and as a result the value of H has changed considerably, but the overall results are the same.

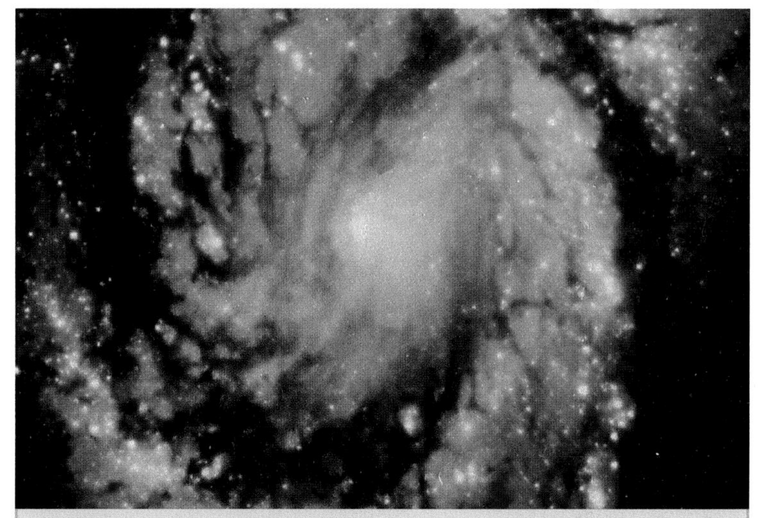

A large spiral galaxy photographed by the Hubble space telescope. Individual stars can be seen in the arms.

CLASSIFICATION OF GALAXIES

Although Hubble's greatest contribution was his discovery of the expansion of the universe, he also made another important contribution. Soon after the study of galaxies began, it became obvious that there were three main types: spirals, ellipticals, and irregulars. Spirals have spiral arms, ellipticals are elliptical in shape and have no arms, and irregulars have no distinguishable shape. The arms in spirals are sometimes tightly wound, and in other cases loosely wound. Most of the galaxies that Hubble examined were spirals.

Hubble decided to classify the galaxies. He subdivided the ellipticals from 0 to 7 according to their roundness, with the roundest being classified E0 and the most elongated being classified E7. Spirals were divided into ordinary spirals and barred spirals (they had a barlike formation through their core). Each of these groups was divided into three classes: *a*, *b*, and *c*, where *a* was the most tightly wound and *c* the most loosely wound. Hubble's original classification has now been extended, but the main features remain.

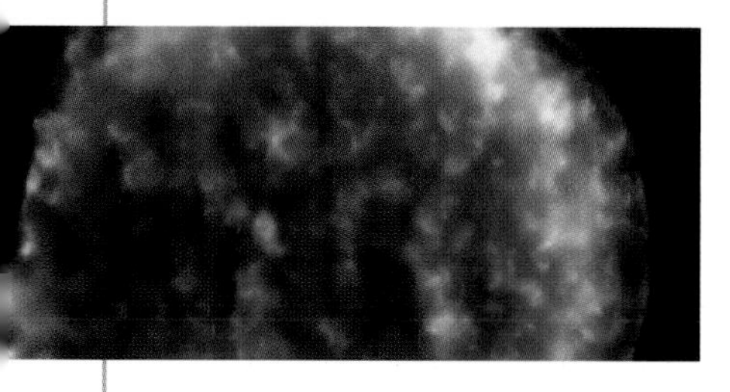

Degenerate Matter and White Dwarfs

In the early 1800s Friedrich Bessel (1784–1846) of Prussia was studying the bright star Sirius when he discovered that its path through the sky had a slight wobble associated with it, rather than a smooth curve as was expected. This meant that it had to have a companion, and since this companion could not be seen, he assumed it was a large planet or a dead star. Twenty-five years later Alvin Clark (1804–87) of the United States was testing a new telescope when he discovered a tiny point of light next to the image of Sirius. Sirius did indeed have a companion, and it was not dark. Walter Adams (1876–1956) of Mount Wilson Observatory obtained its spectrum and determined that its surface temperature was 8,000°K (14,000°F). Astronomers soon determined that it was not much larger than Earth, but thousands of times more dense.

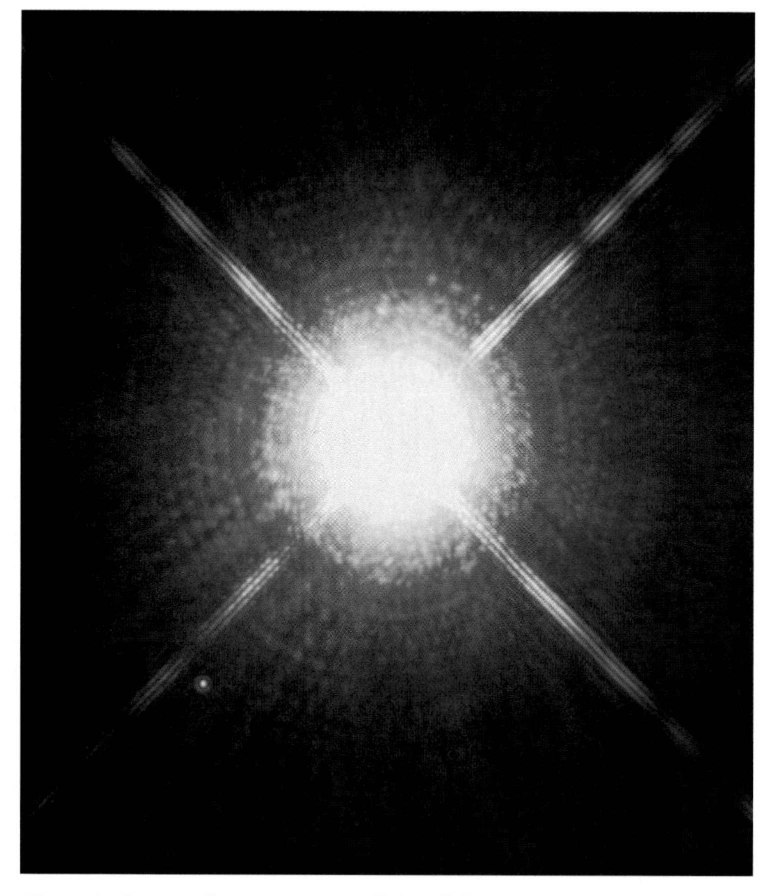

Above: A close-up of our nearest star, Sirius. It has a tiny white dwarf companion. Top left: Image of the remnant of a supernova (an exploding star) that was observed by Tycho in the year 1572.

How could it be so small and dense? It was known that stars died when their thermonuclear furnace went out, but they were not expected to leave an object as dense as this. In 1927 Ralph Fowler (1889–1944) of Cambridge University applied quantum theory to the problem and found that at the extreme temperatures inside a star, atoms would dissociate into a mixture of electrons and nuclei called *degenerate matter.* While in the atom, the electrons and nuclei occupy only about one-trillionth of the total space, so an atom is obviously mostly empty space. But if the electrons and nuclei dissociate and the pressures are sufficiently high, as they are in the interior of a star, the so-called degenerate matter can be compressed to a tiny fraction of its original volume. A star about the size of our Sun, for example, could be compressed to the size of the Earth.

Subrahmanyan Chandrasekhar. Best known for his explanation of white dwarfs, the Indian physicist also made important contributions to black hole physics.

CHANDRASEKHAR'S FINDINGS

Subrahmanyan Chandrasekhar (1910–95) was born in Lahore, India (now part of Pakistan), in 1910. At Cambridge University, where he obtained a PhD in 1933, Chandrasekhar worked on the theory of the small, strange stars that he called white dwarfs.

Chandrasekhar showed that the more massive a star, the smaller it would end up after its collapse, so the smallest white dwarfs were the heaviest. He confirmed Fowler's assertion that the relativistic degenerate matter would be compressed, but he went further and discovered that a star with a mass greater than 1.4 solar masses (times the mass of the Sun) would not stop collapsing at the white dwarf stage. This is now referred to as the *Chandrasekhar limit.*

Chandrasekhar suggested that stars with a mass greater than 1.4 solar masses had to lose some of their mass before they collapsed if they were to end as a white dwarf. And indeed, very massive stars do explode, as supernovae, and in the process they lose considerable mass.

Chandrasekhar presented his ideas on relativistic degenerate matter and the limiting mass to the Royal Society in England and was surprised that they were challenged by Arthur Eddington (1882–1944). Eddington was one of the best-known astrophysicists in the world, and his opinion carried a lot of weight. Arguments on both side of the issue lasted for several years, and Eddington never supported the theory, but in the end Chandrasekhar was proven right. He wrote up his results in a book titled *An Introduction to Stellar Structure.* In the latter part of his life he turned to objects even denser than white dwarfs and became an expert on black holes. He wrote a book titled *The Mathematical Theory of Black Holes* that soon became the bible in the area.

Close-up of a white dwarf star, circled.

Beyond White Dwarfs

What would happen if a supernova explosion did not blow off enough mass before the final collapse of the star, and the star collapsed with a mass greater than 1.4 solar masses? In 1933 Walter Baade (1893–1960) and Fritz Zwicky (1898–1974) of Hale Observatory put forward the idea that the electrons and protons could be squeezed into neutrons (via inverse beta decay) and the remnant would end up as a neutron star. This possibility was proven theoretically in 1939 when Robert Oppenheimer (1904–67) and George Volkoff (1914–2000) of the California Institute of Technology showed that a neutron star could exist and would be stable. And it would be even denser than a white dwarf. Where a teaspoon of material from a white dwarf might weigh five tons, a teaspoon of material from a neutron star would weigh millions of tons. Furthermore, a neutron star would be much smaller than a white dwarf—only about 12 miles (20 km) across. Physicists soon showed that such stars would be spinning very rapidly and have strong magnetic fields. But did neutron stars exist in nature? In 1967 objects called pulsars were discovered, and they were soon shown to be neutron stars.

One of the important results of Oppenheimer and Volkoff's calculation was that there was a limiting mass of about three solar masses for neutron stars. Oppenheimer then went on with Hartland Snyder to show that if the mass was greater than this, the collapse would never stop. In other words, the star did not have a stable end state. Such an object is now referred to as a *black hole.*

STEPHEN HAWKING

Born in Oxford, England, Stephen Hawking (b. 1942) grew up in and around London. Upon graduation from high school he attended Oxford University, where he studied physics and mathematics. From Oxford he went to Cambridge University. A short time after arriving in Cambridge he began to notice that his speech was beginning to slur and he was having difficulty walking. The problem was diagnosed as amyotrophic lateral sclerosis (ALS), known in the United States as Lou Gehrig's disease. Hawking had almost finished his doctoral thesis but was told that he would live only a short time, and he quickly lost interest in finishing it. The disease did not progress as rapidly as expected, however, and Hawking soon

Above: Stephen Hawking, the British cosmologist who made important contributions to cosmology and black hole physics. Top left: In December 2004, this neutron star flared up so brightly it temporarily blinded all X-ray satellites in space. The tremendous blast of energy was caused by a giant flare on the neutron star's surface.

handicaps he has gone on to make a number of remarkable discoveries in physics, and he has also become a prolific author.

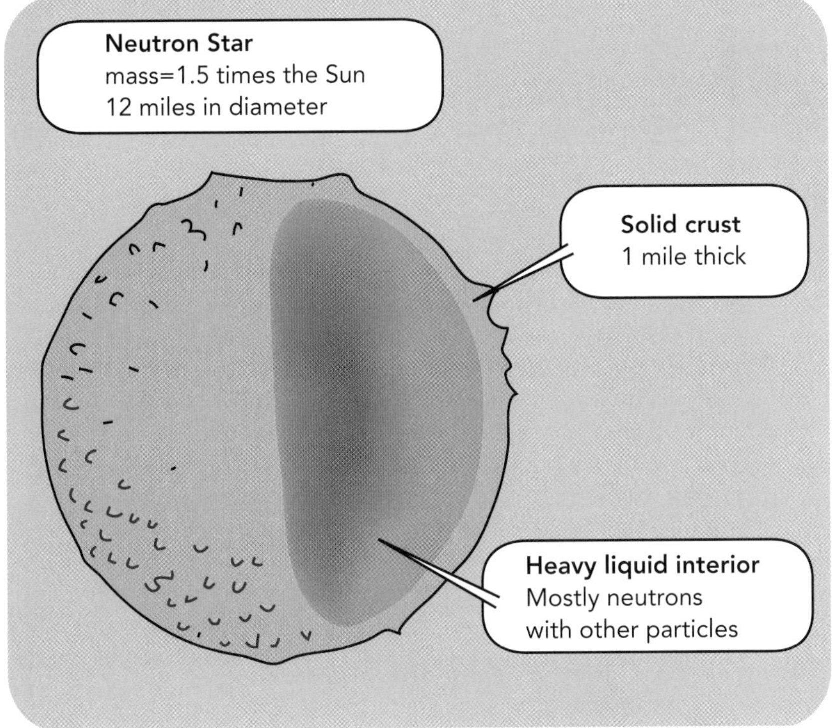

Cross-section of a neutron star, showing the solid crust and liquid interior.

completed his PhD and was later married. Over time, Hawking lost much of his ability to control his muscles and was confined to a wheelchair.

In recent years he has also lost his voice and can no longer move his arms, but he now has a voice machine that gives him an electronic voice. Despite his

MINI BLACK HOLES

One of Hawking's first discoveries was that not all black holes arise in the collapse of massive stars. It is now believed that the universe began as an explosion referred to as the big bang (see page 202). Hawking showed that tiny black holes may have formed in this explosion. If it was inhomogeneous (uneven) during the first fraction of a second, pockets of matter would have been compressed into black holes as a result of the tremendous pressure. They are usually referred to as primordial black holes to distinguish them from stellar collapse black holes.

Stellar collapse black holes are all about the same size, but primordial black holes would have a large range of masses, from tiny ones (mini black holes) about the size of an atom, to huge ones with the mass of a galaxy. There is now evidence, in fact, that many (if not all) galaxies have huge black holes at their core.

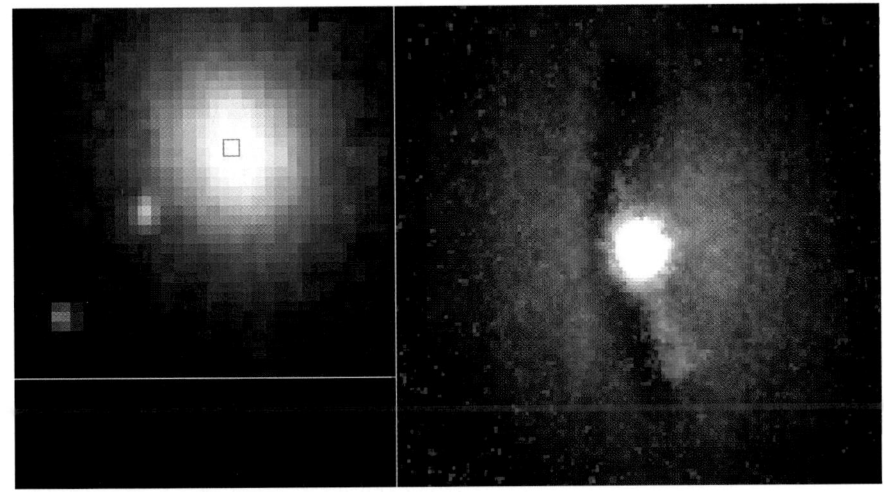

The image of a warped disk flooded with a torrent of ultraviolet light from hot gas trapped around a suspected massive black hole.

Hawking Radiation

Another of Hawking's discoveries is that black holes appear to emit radiation from near their surface. In the early 1970s Jacob Bekenstein (b. 1947) of Princeton University showed that black holes might have a surface temperature greater than zero. This was surprising, since it was assumed that black holes were perfect absorbers and emitted nothing and therefore had to have a surface temperature of zero degrees. Because of this, few people took Bekenstein's result seriously. But Hawking became interested in it and was soon able to explain how it occurred. Applying quantum theory, he showed that the increased temperature is due to a strange type of particle evaporation that was occurring just outside the surface of the black hole.

VIRTUAL PARTICLES

It was believed that black holes were not composed of particles, and it seemed unlikely that they could create particles. Hawking showed that this is not the case. He pointed out that according to relativistic quantum mechanics, particles can be generated in empty space if enough energy is available. This occurs because there is a fuzziness associated with nature at this level, and as a result, particle pairs (for instance, an electron and a positron) can be generated. Their lifetime is very short, however; they are created and quickly come back together and are annihilated, all under the veil of the uncertainty principle. They are referred to as *virtual particles*. Virtual particles of this type are generated just outside the surface of the black hole, and because there are strong forces in this region, they are separated before they can annihilate. Some of the particles fall into the black

Above: Computer illustration of a space shuttle leaving a wormhole in space. Wormholes are created by black holes; they are the curved space associated with the throat of the black hole. Top left: Artist's conception of the Spitzer space telescope pointed toward Earth to receive instructions.

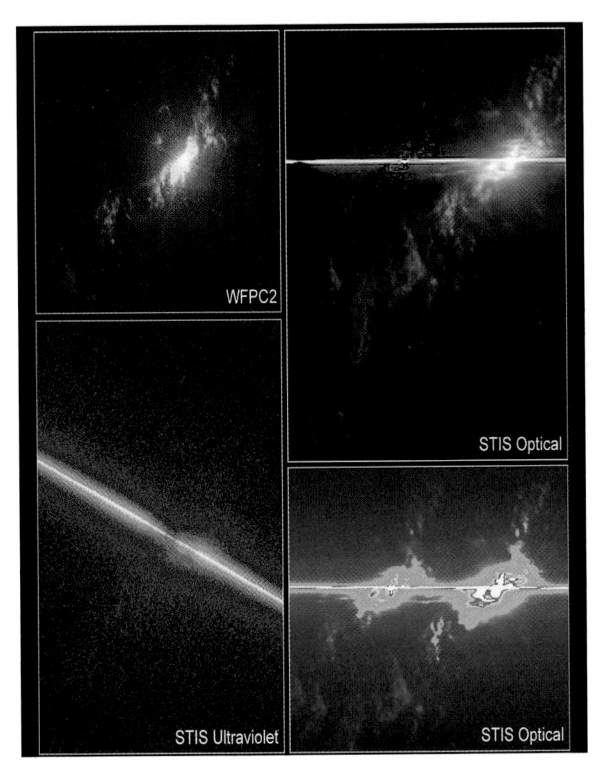

Various images taken of fireworks near a massive black hole in the core of a Seyfert galaxy. Such black holes release enormous amounts of energy.

hole, but many escape, and in the process considerable radiation is generated. From a distance the black hole therefore appears as if it is emitting particles and radiation.

But if the black hole is emitting radiation and particles, it is emitting energy and therefore its mass is decreasing and it is getting smaller. Hawking showed, in fact, that as it gets smaller it radiates more and more energy and it therefore gets hotter and hotter, until it finally explodes.

OTHER CONTRIBUTIONS

Of particular importance, Hawking noticed something curious about the form of the radiation that was being emitted by the black hole. It obeyed the same formula that Planck had derived for the emission of radiation from a heated object. But Planck's formula is a quantum mechanical formula, and black holes are described by Einstein's theory of general relativity, which is a classical theory (these two theories—quantum theory and general relativity—are the two pillars of modern physics, but they are completely separate). A quantum mechanical formula had therefore been derived using Einstein's theory, which meant that there was a link between the two theories, the first that had ever been found. As we will see later, physicists are still trying to bring the two theories together.

In 1971 Hawking, working with Roger Penrose (b. 1931) of England, determined conditions for the existence of "singularities" (places where general relativity breaks down and is no longer valid) in the universe. Also, in collaboration with Jim Hartle (b. 1939) of the University of California at Santa Barbara, he developed an interesting model of the universe that had no boundary. Hawking has also made important contributions to the thermodynamics of black holes.

One of Hawking's greatest successes was his popular book *A Brief History of Time*, which was published in 1988. By 1995 the book had been on the *Sunday Times* best-seller list for 237 weeks, a feat that is recorded in *Guinness World Records*. Since then he has written several other popular books that have also been best sellers.

RECOGNITION AND FAME

Hawking has not received the Nobel Prize, but he has won several prestigious medals, including the Eddington Medal (1975), the Albert Einstein Medal (1979), and the Gold Medal of the Royal Astronomical Society (1985). He is a fan of the television series *Star Trek* and played a brief role in one episode. Interestingly, he played himself, making him the only person on the series ever to do such a thing. Hawking's time at Cambridge University as a PhD student was portrayed in a 2004 BBC movie. His book *A Brief History of Time* was made into a documentary in 1991, starring himself, his family, and a few friends. Hawking has entered popular culture thorough various portrayals on television programs, including *The Simpsons.*

CHAPTER 14

OPEN QUESTIONS AND FRONTIERS OF PHYSICS

Left: An artist's conception of the evolution of the universe. The very early universe is shown to the left. As you move to the right, young galaxies are born (white circles) that eventually develop a spiral or elliptical shape. Top: Two galaxies in collision. Eventually the two will merge into one galaxy, and as they merge new stars will form. Bottom: Artist's illustration of the Chandra observatory spacecraft. In the background is the center of our galaxy.

Physicists have learned a tremendous amount over the last few decades, but for each problem they have solved, new ones have arisen. Most of these problems are related to the smallest objects in the universe, and the universe itself. But there is an object that fits into neither of these cases, and it is one of the most enigmatic entities in the universe; it is called a black hole. There is still much to be learned about these mysterious bodies.

Many problems also remain in relation to our model of the beginning of the universe and its subsequent structure, known as the big bang model. How did the universe begin? How will it end? Does the universe have a boundary? What is its overall structure? And there is also the relationship between the various forces of the universe and the particles to be considered. Einstein searched for years for a unified field theory and was unsuccessful. Many problems, such as the question about whether superstring or M theory is the answer, continue to perplex scientists.

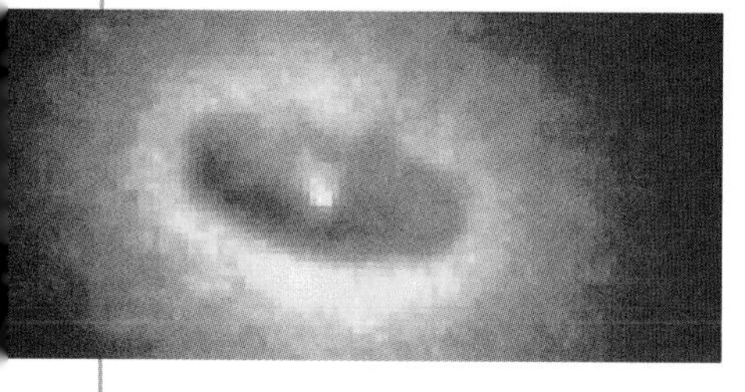

Black Holes

When a star gets old its thermonuclear furnace goes out, and without the outward pressure that this furnace provides, the star collapses in on itself and dies. Depending on its mass, the star can end as a white dwarf, a neutron star, or a black hole. Black holes are quite different from white dwarfs or neutron stars. One of the major differences is that black holes do not have a solid surface. If you approached a black hole, you would find that you could pass right through its surface. Most of it is empty space; the mass of the collapsed star is at the center of the black hole.

Whether a collapsing star becomes a black hole depends on what is called its *gravitational radius*. If all the mass of a star collapses inside this radius, the object will become a black hole with a radius of a few miles.

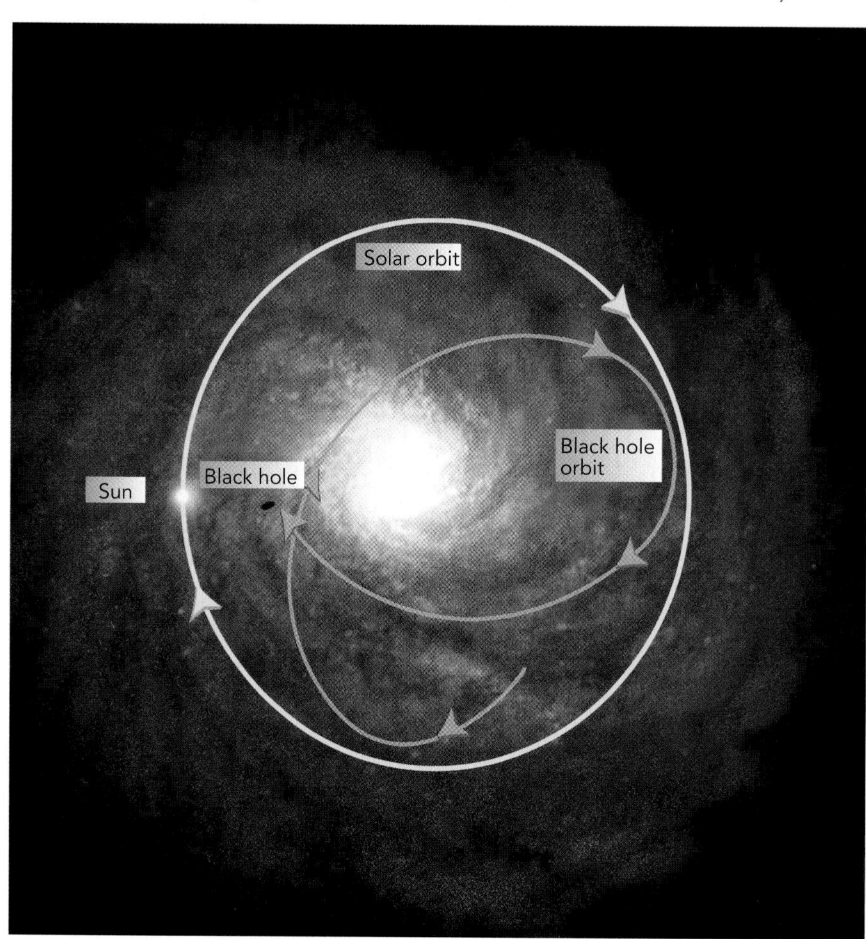

Above: The orbit of a black hole that is believed to have been catapulted from a supernova explosion that occurred many years ago. Both our Sun and the black hole orbit the center of the Milky Way galaxy. Top left: Ring around a suspected black hole in a distant galaxy. Indirect evidence of black holes comes from studying the effects on their surroundings.

STRUCTURE OF A BLACK HOLE

The black surface we see from a distance is called the *event horizon* of the black hole. At the center of this event horizon is the *singularity*; or the remnant of the star (all the mass of the star is here). Anything that gets close to a black hole is pulled into it. Once something passes through the event horizon, there is no way it can ever get out.

Black holes have many interesting and exotic properties. Time, for example, slows down as you approach the event horizon of a black hole. But this slowing is noticeable only to someone observing it from a distance. If an observer A, for example, approaches a black hole with a clock and another observer, B, is watching him from a distance, observer B will see A's clock running slower and slower. In theory, A's clock as observed by B, will stop as A passes through the event horizon. Things are different, however, for observer A watching his own clock. Time passes normally for him, and when he reaches the event horizon, he passes through it to the interior of the black hole.

One of the most interesting properties of a black hole is the curved space near it. It resembles a funnel and is referred to as the *throat* of the black hole. Such objects are now frequently referred to as *wormholes* in space.

BLACK HOLE CANDIDATES

We are unable to observe black holes directly, so we have to look for indirect evidence of their existence. Cosmologists study the skies seeking good candidates for black holes. Consider there is a double-star system in which one of the stars has collapsed to a black hole. If the black hole pulled matter in from its companion star, X-rays would be emitted just before the matter passed through the event horizon. A strong X-ray source in a double star system where one of the stars was not visible would therefore quite possibly be a black hole. And indeed, several candidates of this type have been found. One of the best known is called CYG X-1; it is in the constellation Cygnus, and the invisible component of the system has been shown to have a mass greater than three solar masses.

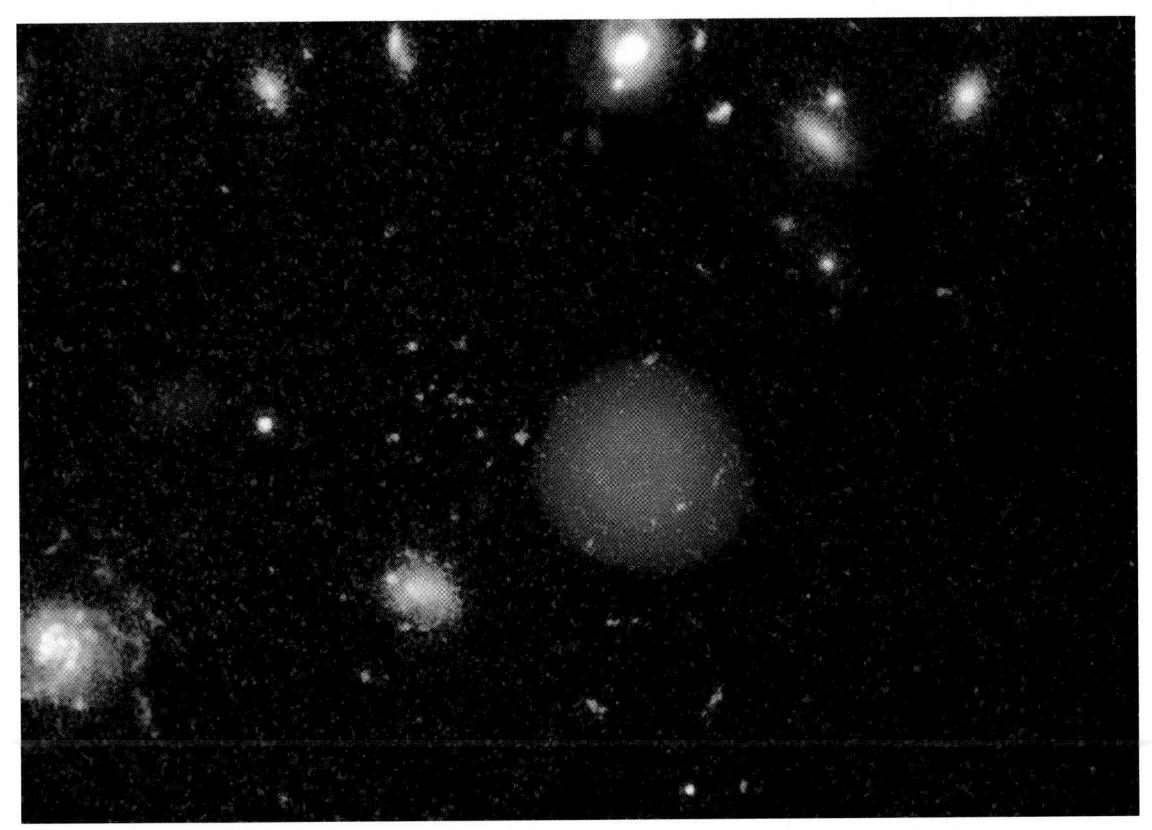

X-rays are being emitted from heated material that is falling into a black hole in the center of this photo.

The Big Bang Model of the Universe

The big bang model predicts an age for the universe of about 15 billion years, which means that the explosion that created the universe occurred 15 billion years ago. One of the first questions we might ask about this explosion is whether it occurred in an infinite space. It seems that it did not. Rather, the explosion created space. Before it, there was no space. Another interesting question involves the galaxies. When we look out into space, the galaxies around us all appear to be receding from us. Does this mean that we are at the center of the universe? Again, the answer is no. Regardless of where you are in the universe, all galaxies will appear to be moving away from you. In effect, the galaxies are all moving away from one

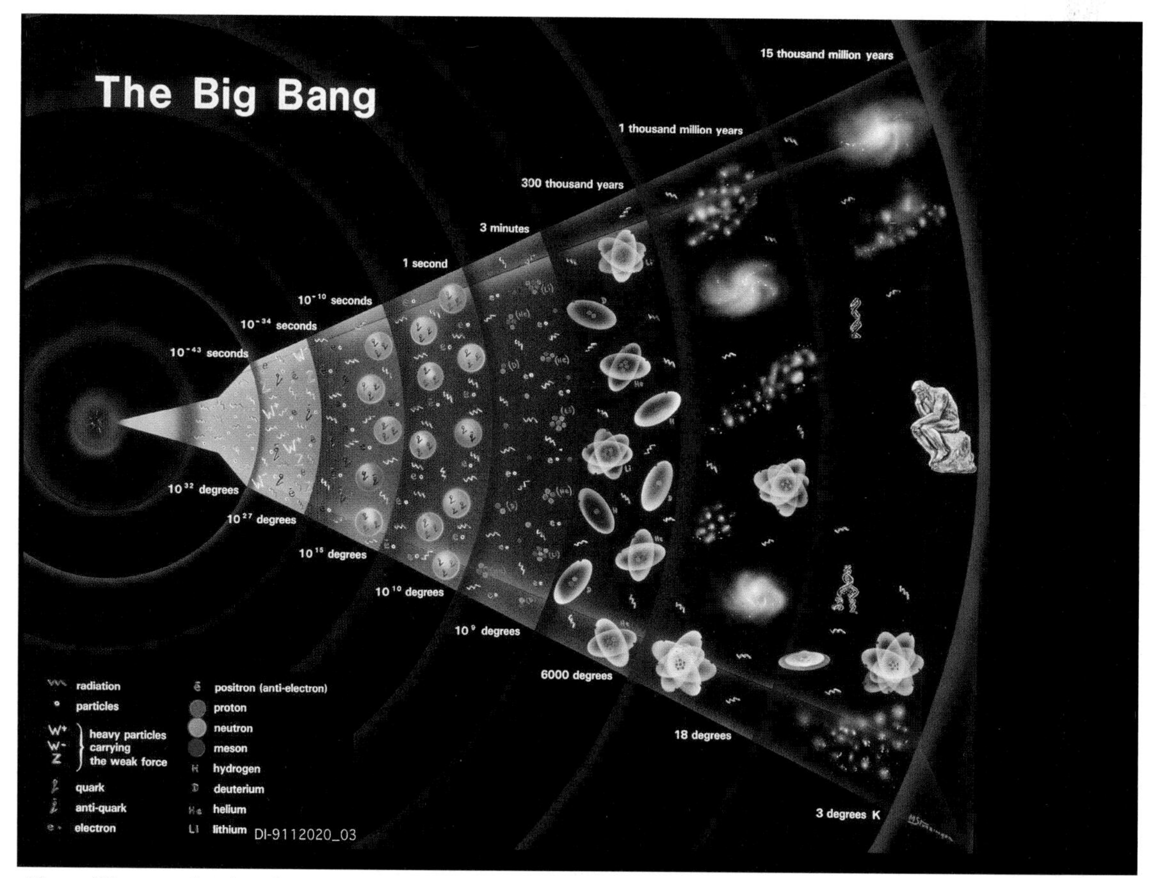

Above: Diagram showing the various stages of evolution of the universe from the big bang to the present time. The big bang is shown at the center of the circles. During its early stages the universe went through several eras, ending with that in which galaxies were formed. Top left: This NASA computer rendering is a simulation of a distant cluster of galaxies.

another—the space between them is expanding. The galaxies themselves are not changing in size. The best way to visualize this is to glue small circles on a balloon, then blow it up. The distance between each of the spots increases, but the spots stay the same.

And while the universe has no end, it does have a "horizon." This is the point where the velocity of the galaxies relative to us is equal to the velocity of light.

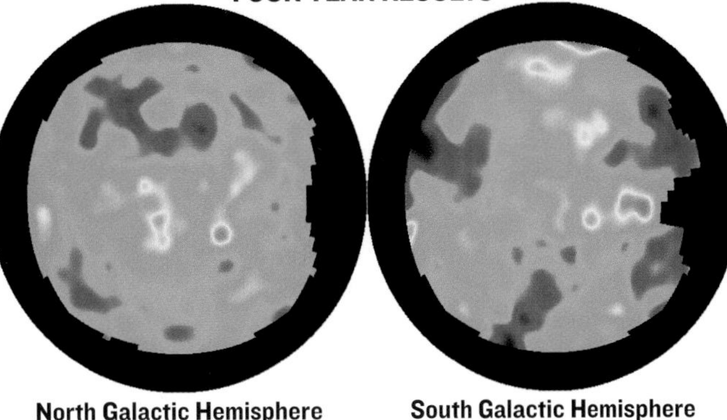

A map of the cosmic background radiation as obtained from the satellite COBE. It shows the variation in intensity in various directions. This radiation is believed to be a remnant of the big bang explosion.

THE COSMIC BACKGROUND RADIATION

George Gamow (1904–68) defected from the Soviet Union in 1934. While working at George Washington University in 1948, he and Ralph Alpher (b. 1921) predicted that radiation should have been produced in the early universe. According to their calculations the radiation would have been extremely hot, but it would have gradually cooled and would now be only a few degrees Kelvin.

Fifteen years later Robert Dicke (1916–77) of Princeton University arrived at the same conclusion (he had not heard of Gamow's prediction). Along with James Peebles (b. 1935) he showed that the radiation should have a temperature of about 3°K. Dicke encouraged two colleagues, P. G. Roll and T. D. Wilkinson (1935–2002), to look for it, and they began building the appropriate equipment. Unknown to them, however, two scientists at Bell Labs in New Jersey, Arno Penzias (b. 1933) and Robert Wilson (b. 1936), had already found it. They were working with a large radio telescope in Holmdel, New Jersey, when they discovered a strange noise that appeared to be coming from the telescope. Checking everything carefully, they found it was not coming from the telescope, but from space. Dicke and Peebles were told about Penzias and Wilson's research, and visited them at Holmdel. They soon realized that the "noise" was the radiation they had predicted.

But only one point in the radiation curve had been obtained. Other points were needed to ensure that this was indeed the cosmic background radiation, and several were soon obtained. The complete curve was achieved by the satellite Cosmic Background Explorer (COBE), which was launched in 1989. This removed all doubt—what they had observed was the cosmic background radiation predicted by Gamow and by Dicke and Peebles.

IS THE UNIVERSE OPEN OR CLOSED?

One of the major problems in cosmology is the future of the universe. In particular, is the universe open or closed? If it is open it will continue expanding forever. If it is closed it will eventually stop expanding and collapse back on itself. And according to Russian physicist Alexander Friedmann (1888–1925), this is determined by comparing the average density of matter in the universe to the critical density, which is approximately 6×10^{-30} gms/cm^3. If the average density is greater than this, the universe is closed, and if it is less, it is open. It might seem that it would be relatively easy to measure the average density of matter, but it is not. The problem is that there is a lot of matter we cannot see. Galaxies and visible matter give an average density of only about 1 percent of the critical density.

Big Bang Difficulties and Dark Matter

The big bang theory has been very successful over the past few decades. In particular it predicts the cosmic background radiation and its temperature, and it predicts the abundance of some of the lighter elements. Nevertheless, it does have several problems. One, called the *flatness problem,* was pointed out by Dicke in 1969. He showed that the universe had to be flat—in other words, it has to have a density equal to the critical density, otherwise it would have collapsed in on itself long ago, or expanded rapidly to emptiness.

A second dilemma is called the *horizon problem.* Charles Misner of the University of Maryland showed in 1969 that opposite ends of the universe are separated by too great a distance to ever have been in communication. Yet the radiation on one edge is at a temperature of 3°K, and at the opposite edge it is also at 3°K. This seems impossible. Another problem, called the *galaxy* or *structure problem,* is concerned with discovering how the galaxies were formed and with the strange overall structure to our universe, in which large holes and bubbles are distributed throughout the galaxies.

INFLATION THEORY

Several of the problems of the big bang theory were apparently solved by an idea put forward in 1980 by Alan Guth (b.1947) of Massachusetts Institute of Technology. He showed that if a sudden inflation, or rapid increase in the expansion rate, of the universe occurred in the first fraction of a second of its existence, many of the problems could be explained. In particular, his theory explains the flatness and horizon problems, and to some degree, the galaxy problem. It also explains where the energy for the expansion comes from. Guth's theory was modified a year later, but it still has some difficulties and is not accepted by everyone. One of its major predictions is that the universe is flat.

Above: American cosmologist Alan Guth. He put forward the inflation theory, an idea that assumes a sudden expansion (much faster than the normal expansion) occurred in the early universe. Top left: A spiral galaxy referred to as M31. It is one of the Milky Way's nearest neighbors in intergalactic space. It and other galaxies are filled with dark matter.

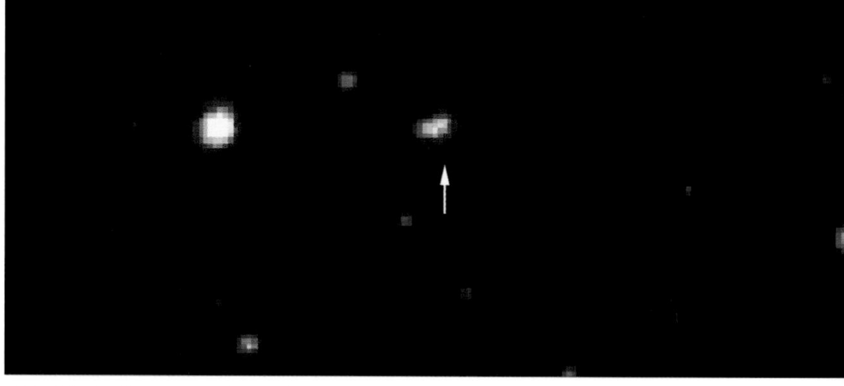

This NASA/European Space Agency Hubble photograph shows a dark matter object—the red dwarf star indicated by the arrow.

DARK MATTER

How could Guth predict a flat universe when all the visible matter in the universe adds up to less than 1 percent of the mass needed to make a flat universe? For years scientists have known that there is a considerable amount of mass in the universe that cannot be seen directly, though we have gravitational evidence that it is there. We now refer to the matter that we cannot see as *dark matter.*

However, if the universe is flat, and 1 percent of its mass is visible, then 99 percent is dark matter. Also, theoretical arguments now show that ordinary matter can make up only 10 percent of the critical density. This means that most of the dark matter has to be in a strange, exotic form rather than ordinary matter.

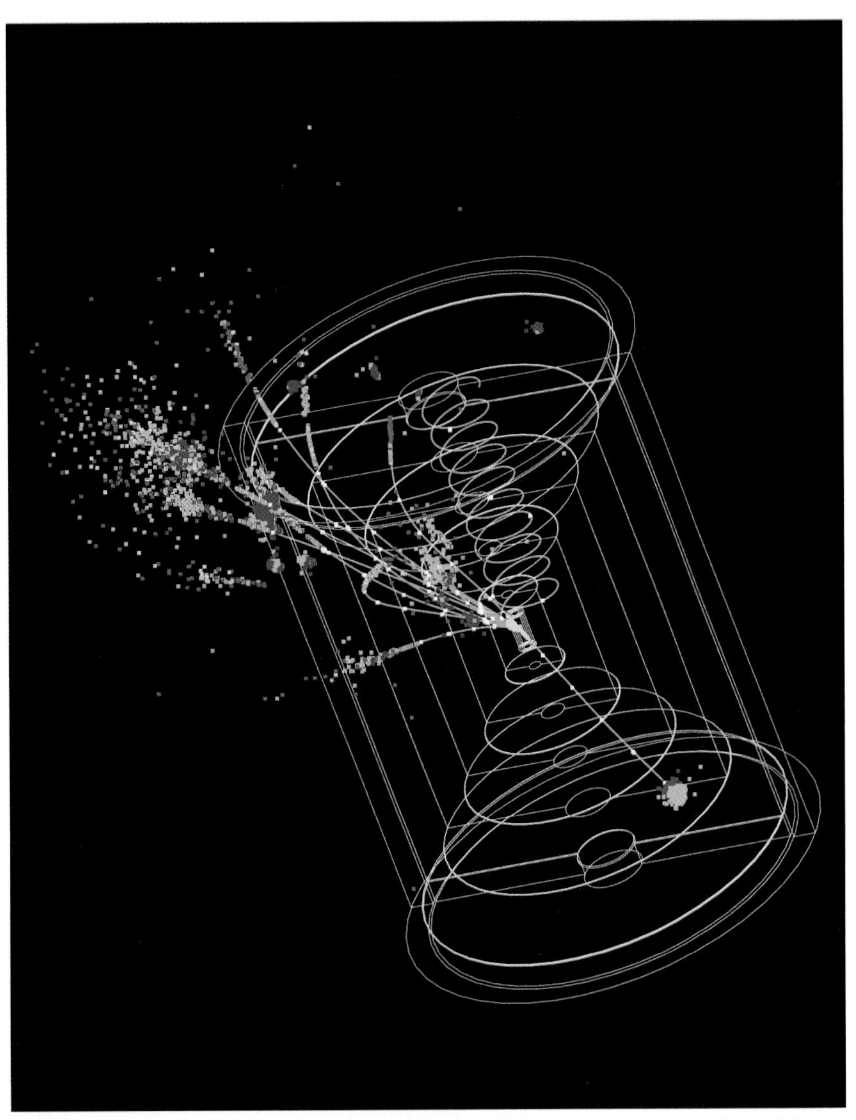

Many types of particles have been suggested as candidates for dark matter. One of them is known as neutralino. This is a simulation of the "signature" that a neutralino is expected to leave in a detector.

THE ACCELERATING UNIVERSE AND DARK ENERGY

For years it was assumed that the expansion of the universe was decelerating. This was based on the belief that the mutual gravitational pull of the matter in the universe would have an effect. Each galaxy pulls on the others, and even if they are moving apart, this mutual attraction should eventually slow them down. To the surprise of astronomers, however, observations in the late 1990s began to show that the universe was not decelerating—it was accelerating. This seemed impossible. But several studies confirmed the discovery, and soon the idea was generally accepted.

But if the expansion of the universe was accelerating, what was causing it? Astronomers now believe that it might be caused by "dark energy." And surprisingly, much of the energy of the universe may be in this form, possibly over 70 percent.

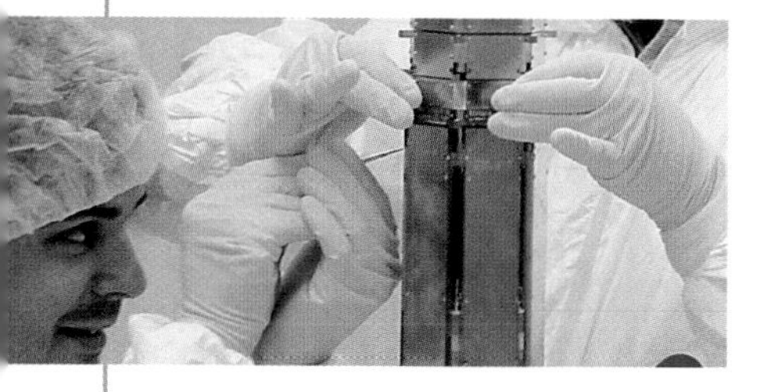

Unified Field Theory

When Einstein completed his general theory of relativity in 1916, only two fundamental forces, or fields of nature were known: the gravitational field and the electromagnetic field. His theory explained gravitation, but Einstein was sure the two fields were related, and an extension of general relativity should be able to explain both fields. Furthermore, he believed this extension could also explain the elementary particles in a more fundamental way. We now refer to this extension as a unified field theory.

In 1929 Einstein introduced a unified field theory. There was so much excitement about the new theory that hundreds of reporters lined up outside his house. But Einstein soon found a flaw in the theory and discarded it.

INCREASED COMPLEXITY

Over the next 30 years Einstein continued to search for a unified theory, but he never found one. In the meantime the problem became considerably more complex. When Einstein first began working on it, only the two fields, the electromagnetic and gravitational, and only two elementary particles, the electron and proton, were known. Over the

Above: Einstein standing before a blackboard. Einstein spent the last 30 years of his life searching in vain for a unified field theory. Top left: In an underground laboratory, about a half a mile from the surface, scientists search for evidence of dark matter. Underground laboratories are needed for searches such as this to avoid unwanted interference from other particles from space such as cosmic rays.

Pakistani physicist Abdus Salam received the Nobel Prize for his unification of the electromagnetic and weak nuclear forces.

AN UNSOLVABLE PROBLEM?

A successful unified field theory would have to unify the four fields: gravitation, electromagnetism, the strong nuclear force, and the weak nuclear force. Because gravity was so different from the other fields, physicists began looking into the possibility of unifying the other three fields. Steven Weinberg (b. 1933), who was at MIT, and independently, Abdus Salam (1926–96) of Pakistan, showed that the weak nuclear theory could be unified with the electromagnetic theory; the new theory was called electroweak, or EW theory. The next step was to unify the strong nuclear field with EW theory. The first to attempt this were Howard Georgi (b. 1947) and Sheldon Glashow (b. 1932) of Harvard; in 1973 they introduced what is called grand unified theory. It was a five-dimensional theory with five basic particles (three quarks, the electron, and the positron). A new particle, called the X-particle, was also introduced; it changed a lepton into a quark and vice versa. The Georgi-Glashow theory is now known to be incorrect, but several other grand unified theories (GUTs) have been put forward. The outlook for them, however, is bleak at the present time.

next few years, however, two more fields of nature were found, and the number of elementary particles increased dramatically. And with the development of quantum mechanics in the late 1920s, there was considerable uncertainty as to how quantum theory would fit in a unified field theory that was based on general relativity. The two theories were completely different. A quantized version of general relativity would be needed if the two theories were to be brought together. Over the years many attempts have been made to produce such a theory, but none has been successful.

American physicist Sheldon Glashow was involved in the unification of the electromagnetic and weak nuclear forces and later introduced a theory that led to quantum chromodynamics (QCD).

Superstring Theory

For a true unified field theory, or, as many prefer to call it, a theory of everything (TOE), the gravitational field would have to be included. But gravity is so different from the other fields that it presents a serious challenge. To many physicists the best approach was an entirely new theory, and the beginnings of one came in the early 1970s. The idea was that particles and forces were made up of tiny strings. In essence, strings rather than point particles would be the elementary units of the universe. Few people were interested in the idea initially, as it seemed far too radical.

Above: Artist's conception of a hypothetical spacecraft that uses "negative energy" to produce hyperfast speeds that would one day allow us to visit distant stars. Top left: Particle accelerators, which propel electrons and anti-electrons to nearly the speed of light, are either linear or circular. The longest linear accelerator in the world, at the Stanford Linear Accelerator Center in California, extends 2 miles (3 km).

Then in 1979 John Schwarz (b. 1941) of the California Institute of Technology and Michael Green (b. 1946) of Queen Mary College teamed up to work on the theory. At the time there were still serious problems with it. By 1981, however, they had overcome most of the problems and had a theory that appeared to unify all four fields of nature, including gravity. Furthermore, it appeared to predict elementary particles; but it still had problems, so in 1984 they extended it to 10 dimensions. However, with the new theory, which is usually referred to as superstring theory, the strings became much smaller. Previously they had been about 10^{-13} cms—about the size of a proton. Now they were about 10^{-33} cm long, or one hundred billion billion times smaller than the nucleus.

MULTIPLE THEORIES

Despite the tiny size of the string, the new theory caught on, and hundreds of papers soon appeared on the subject. The strings themselves could be open or closed, and it was assumed that they vibrated with various frequencies. In some cases waves

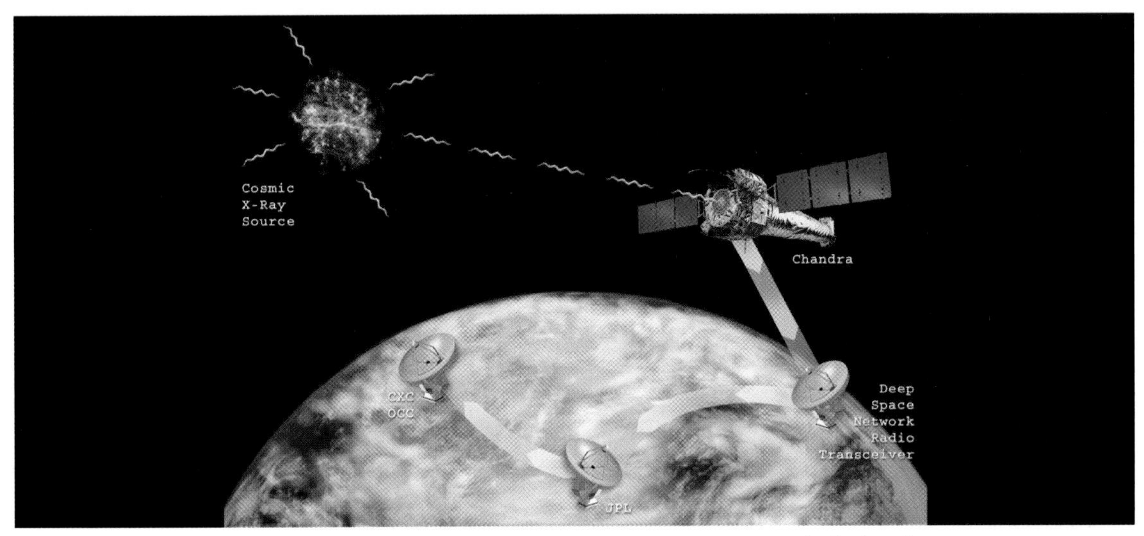

Artist's depiction of the Chandra satellite detecting X-rays from a distant source and relaying the information to radio telescopes on Earth. Below: A computer-generated representation of superstrings in a multidimensional space-time.

moved along the strings, and in others they were standing, or stationary waves. These waves were similar to the waves created on a violin string.

During the 1980s considerable work was done on superstring theory, and by the early 1990s there were five different and independent superstring theories. This presents a dilemma to string theorists. Which theory is correct? The ultimate string theory or TOE obviously can not be the combination of five separate theories. The only alternative is that one of the theories is correct and the other four incorrect, or perhaps all five are somehow equivalent.

To everyone's surprise, Edward Witten (b. 1951) showed in 1995 that all five theories are related. They are, in effect, just five different ways of expressing the same physics. Witten called the new theory M theory, but no one is quite sure what the M stands for. Various candidates are "mother theory," "membrane theory," and "matrix theory." In developing M theory, however, Witten had to add a dimension, so the new theory has 11 dimensions.

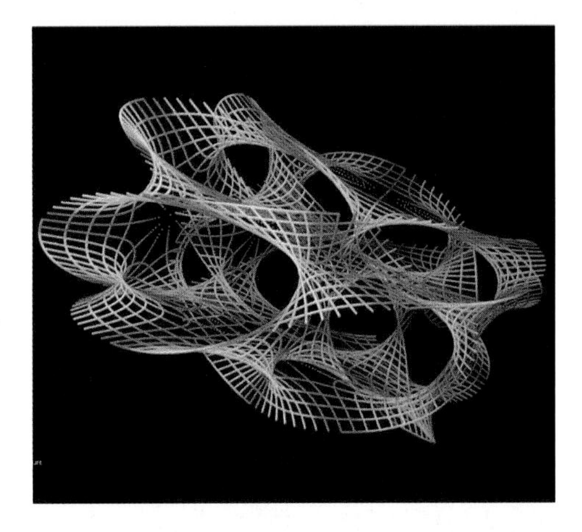

BRANES

One consequence of Witten's theory is that the one-dimensional strings of superstring theory can become two-dimensional membranes. Physicists immediately began asking if even higher dimensional membranes are possible. Paul Townsend of Cambridge University introduced the terminology *branes* to describe these membranes that can have any dimension up to nine. He referred to them as p-branes, where p refers to the dimension. Simply, a brane is an extension of a string to a higher dimension. Taken together, the branes give us a "brane world," in which three of the forces of nature are trapped on branes. The fourth force, gravity, would be free to roam in the region around the branes.

GLOSSARY

A

ABSOLUTE TEMPERATURE The temperature scale with zero as lowest possible temperature; in practice, it is the Celsius (centigrade) scale plus 273.

AMPLITUDE For a vibration around an equilibrium position, the distance from the equilibrium position to the extreme of motion.

APHELION The point in an orbit of a celestial body when it is farthest from the Sun.

B

BIG BANG A cosmology theory that assumes the universe began as a gigantic explosion.

BLACKBODY A hypothetical body that absorbs and emits all radiation that falls on it.

BLACK HOLE A celestial object with a gravitational field so strong that light cannot escape from it.

C

CALCULUS A branch of mathematics that deals with small changes in such things as position, velocity, and time.

CENTER OF GRAVITY The balancing point between two or more masses, or within a single mass.

CONDUCTION The transference of heat by means of contact between neighboring molecules.

CONVECTION Transfer of heat from one place to another by actual motion of the hot material.

COSMIC BACKGROUND RADIATION Thermal blackbody radiation that permeates the universe; it has a temperature of 2.7 degrees Kelvin.

D

DARK MATTER Matter in the universe inferred to exist because of its gravitational attraction rather than by direct observation.

DIFFRACTION The slight bending of light rays that occurs as they pass very closely by opaque objects or through small openings.

DISPERSION The spreading out of light into its component colors or frequencies.

DOPPLER EFFECT A change in wavelength that occurs when a body emitting a wave is either approaching or receding from an observer.

E

ELECTRIC FIELD Lines assumed to be emanating from a positive or negative charge.

ELECTROMAGNETISM The study of the combined effects of electricity and magnetism.

ELECTROSTATICS The branch of physics dealing with the properties of electricity at rest.

ELEMENTARY PARTICLE A basic particle out of which all other particles, or matter, is made.

ENTROPY A measure of the disorder of a system; numerically expressed as heat divided by absolute temperature.

F

FISSION The breaking apart or splitting of nuclei into two or more smaller parts.

FREQUENCY Number of waves passing a given point per second.

G

GALAXY A large system of stars, usually containing millions to hundreds of billions of stars, with gas and dust also frequently present.

GAMMA RAY Electromagnetic radiation with the highest energy and frequency.

H

HALF-LIFE Time required for one half of the atoms in a radioactive sample to disintegrate.

I

IDEAL GAS A hypothetical gas that obeys the gas laws perfectly at all temperatures and pressures.

INERTIA Resistance to a change in motion.

ION A particle with a positive or negative charge.

ISOTOPE One of the several forms of a chemical element that have the same number of protons in the nucleus but differ in the number of neutrons and thus differ in atomic weight.

K

KELVIN SCALE See *absolute temperature.*

KINETIC ENERGY The energy of motion.

L

LASER A device that produces coherent light with waves synchronized.

LIGHT YEAR The distance light travels in one year.

M

MASS A measure of the quantity of matter in a body.

MOLECULE A combination of atoms, bound together by an electromagnetic force, that forms the smallest unit of a compound.

MOMENTUM The product of the mass times the velocity of an object.

N

NEUTRON The neutral particle of the atomic nucleus.

NUCLEAR FORCE The force that holds the particles of the nucleus together.

NUCLEAR REACTION A reaction involving nuclear particles.

NUCLEUS The central part of the atom, which contains protons and neutrons.

O

ORBIT The path an object traces in its revolution around another while under the influence of gravity.

P

POSITRON A positive electron, or antiparticle of the electron.

Q

PROTON An elementary particle found in the nucleus of the atom.

QUANTUM A "bundle," or discrete amount of a quantity.

QUANTUM NUMBER Whole numbers or half numbers that label the state of a system, such as an electron in an atom.

QUANTUM THEORY Based on the idea of a quantum, the theory that energy is assumed to come in small units.

R

RADIATION The emission of energy in the form of waves. Also, particles emitted by radioactive substances.

RADIOACTIVITY The disintegration of atomic nuclei with the release of highly energetic radiation or particles.

REFRACTION The bending of light as it passes into a more or less dense translucent material.

RETROGRADE MOTION An apparent motion of a planet in space opposite to its usual pattern of movement or to that of most of the other planets in its solar system.

S

SCALAR A quantity such as temperature or volume that has no direction, only magnitude.

SPECTRUM The array of electromagnetic radiations spread out in order of wavelength.

V

VECTOR A quantity that has both magnitude and direction.

VELOCITY A measure of the rate of change in position with respect to time, involving speed and direction.

W

WAVELENGTH The distance between equivalent points of a wave, for instance from trough to trough or crest to crest.

WORMHOLE A space-time tunnel associated with a black hole.

X

X-RAY Energetic radiations between those of gamma and ultraviolet rays.

FURTHER READING

BOOKS

Aczel, Amir. *God's Equation.* New York: Four Walls Eight Windows, 1999.

Asimov, Isaac. *The History of Physics.* New York: Walker & Co., 1966.

Bodanis, David. *$E = mc^2$.* New York: Berkley Books, 2000.

Boorstin, Daniel. *The Discoverers.* New York: Random House, 1983.

Brian, Denis. *Einstein: A Life.* New York: Wiley, 1986.

Christianson, Gale. *In the Presence of the Creator: Isaac Newton and His Times.* New York: Free Press, 1984.

Crease, Robert, and Charles Mann. *The Second Creation.* New York: McMillan, 1986.

———. *The Whole Shebang.* New York: Simon and Schuster, 1997.

Ferris, Timothy. *Coming of Age in the Milky Way.* New York: Doubleday, 1988.

Gardiner, Martin. *The Relativity Explosion.* New York: Vintage Books, 1976.

Gilmore, Robert. *The Wizard of Quarks.* New York: Copernicus Books, 2001.

Greene, Brian. *The Elegant Universe.* New York: W. W. Norton, 1999.

Gribbin, John. *Almost Everyone's Guide to Science: The Universe, Life and Everything.* New Haven, CT: Yale University Press, 1998.

Hawking, Stephen. *A Brief History of Time.* New York: Bantam, 1988.

Hewitt, Paul. *Conceptual Physics.* Boston: Addison Wesley, 2005.

March, Robert. *Physics for Poets.* New York: McGraw-Hill, 1978.

Overbye, Dennis. *Einstein in Love.* New York: Penguin, 2001.

Parker, Barry. *Albert Einstein's Vision: Remarkable Discoveries that Shaped Modern Science.* Amherst, NY: Prometheus Books, 2004.

———. *Einstein's Brainchild: Relativity Made Relatively Easy.* Amherst, N.Y.: Prometheus Books, 2000.

———. *Einstein: The Passions of a Scientist.* Amherst, NY: Prometheus Books, 2003.

———. *Search for a Supertheory.* New York: Plenum, 1987.

Rhodes, Richard. *The Making of the Atomic Bomb.* New York: Simon & Schuster, 1986.

Trefil, James. *Physics as a Liberal Art.* New York: Pergamon Press, 1978.

Wolf, Fred. *Taking the Quantum Leap.* San Francisco: Harper & Row, 1981.

WEB SITES

American Physical Society
www.aps.org

American Scientist Online
www.americanscientist.org

Boston Museum of Science/Inventor's Toolbox
www.mos.org/sln/Leonardo/InventorsToolbox.html

Einstein's Big Idea
www.pbs.org/wgbh/nova/einstein

Georgia State University Department of Physics and Astronomy
hyperphysics.phy-astr.gsu.edu/hbase/hframe.html

Physics Central
www.physicscentral.com

The Physics Classroom
www.physicsclassroom.com

Physics to Go
www.physicstogo.org

A Science Odyssey: People and Discoveries
www.pbs.org/wgbh/aso

AT THE SMITHSONIAN

The Smithsonian Institution runs a large number of projects in astrophysics and planetary physics in various locations around the world, ranging from the submillimeter telescope project in Antarctica to the submillimeter array of radio telescopes at Mauna Kea in Hawaii, and the Chandra Observatory, a satellite launched in 1999.

The National Air and Space Museum in Washington, D.C., houses the largest collection of historic aircraft and spacecraft in the world. It is an important research center, where all aspects of aviation and space flight are explored, as are planetary science and terrestrial geology and geophysics.

Online exhibitions (www.nasm.si.edu/exhibitions) as diverse as *The Wright Brothers and the Invention of the Aerial Age* and *GPS: A New Constellation* explore the many ways that physics influences our lives.

CENTER FOR EARTH AND PLANETARY STUDIES

The Center for Earth and Planetary studies (CEPS) is a research unit within the National Air and Space Museum. Scientists on staff at CEPS research topics in planetary science, geology, and geophysics, administer the collection, and perform public outreach.

ALBERT EINSTEIN PLANETARIUM

Located in Washington, D.C., the planetarium operates with the latest in digital technology. The central instrument of this planetarium, a Zeiss projector, is one of the best in the world. It allows the sensation of zooming through the cosmos, surrounded by stars and galaxies. A new feature is a tour of the solar system, out past the Milky Way to the edge of the universe and back.

SMITHSONIAN ASTROPHYSICAL OBSERVATORY

The Smithsonian Astrophysical Observatory (SAO) is a research institute operated in conjunction with Harvard University. The home facility is located at the Harvard-Smithsonian Center for Astrophysics (CfA) in Cambridge, Massachusetts. More than 300 scientists are involved in projects in astronomy, astrophysics, earth and space science, and science education at the center's various facilities around the world.

CHANDRA OBSERVATORY

The Chandra X-ray Observatory was launched on July 23, 1999. Originally known as the Advanced X-ray Astronomy Facility (AXAF), it was renamed in honor of the well-known astrophysicist Subrahmanyan Chandrasekhar. Its prime mission is to observe X-rays in space. The control center for the satellite is in Cambridge, Massachusetts. You can view Chandra images at http://chandra.harvard.edu/resources/misc/special_features.html.

ANTARCTIC SUBMILLIMETER TELESCOPE

Antarctica is an ideal place for submillimeter observations because of its cold, dry atmosphere. The Smithsonian's telescope is located at the Amundsen-Scott South Pole station, and is currently focused on observations of submillimeter emissions from gas clouds in the Milky Way and other galaxies.

WHIPPLE OBSERVATORY

Located in Amado, Arizona, the major instrument of the Whipple Observatory is the Multiple Mirror Telescope (MMT), which is a joint venture between the Smithsonian and the University of Arizona. A 6.5-meter telescope, the MMT is used to study extrasolar planets, stellar motion, and galaxies, among other things.

VERITAS

Gamma rays have been of considerable interest to astronomers in recent years. VERITAS (Very Energetic Radiation Imaging Telescope Array System) is designed to study these rays, and should be operational in late 2006. It consists of four 12-meter reflectors and a telescope imaging system, and will be used to study supernovae and active galactic nuclei. NASA will launch a gamma ray satellite in 2007 and VERITAS will operate in conjunction with it. VERITAS is a joint project of the Smithsonian and ten leading universities.

INDEX

S

T

U

V

W

X

Y

Z

ACKNOWLEDGMENTS & PICTURE CREDITS

The publisher wishes to thank consultant John Balbach, Physics Department, George Washington University; Lynn Carter, Center for Earth and Planetary Studies, National Air and Space Museum; Ellen Nanney, Senior Brand Manager with Smithsonian Business Ventures; Katie Mann and Carolyn Gleason with Smithsonian Business Ventures; Collins Reference executive editor Donna Sanzone, editor Lisa Hacken, and editorial assistant Stephanie Meyers; Hydra Publishing president Sean Moore, publishing director Karen Prince, editor Molly Morrison, copy editor Glenn Novak, editorial director Aaron Murray, art director Brian MacMullen, designers Erika Lubowicki, Ken Crossland, Eunho Lee, Pleum Chenaphun, Gus Yoo, and La Tricia Watford, editors Marcel Brousseau, Ward Calhoun, Suzanne Lander, Rachael Lanicci, Michael Smith, Liz Mechem, and Amber Rose; picture researcher Ben DeWalt; indexer Jessie Shiers; Wendy Glassmire of the National Geographic Society; Harriet Mendlowitz of Photo Researchers, Inc.

The following abbreviations are used: PR–Photo Researchers, Inc.; SPL–Science Photo Library; JI–© 2006 Jupiterimages Corporation; SS–Shutterstock; IO-Index Open; IS–iStockphoto.com; BS–Big Stock Photos; NSF–National Science Foundation; NASA–National Aeronautics and Space Administration; GSFC–Goddard Space Flight Center; MSFC– Marshall Space Flight Center; JPL– Jet Propulsion Laboratory; STScI– Space Telescope Science Intitution; SLAC–Stanford Linear Accelerator Center; SIL–Smithsonian Institute Libraries; AP–Associated Press; LOC–Library of Congress; NGIC–National Geographic Image Collection; FS–Fotosearch; AIP–American Institute of Physics

PICTURE CREDITS

(t=top; b=bottom; l=left; r=right; c=center)

Introduction
IV SS/Jurgen Ziewe **VI** SS/Tom Hirtreiter **1t** SS/Andrea Danti **1b** IO **2** SS/Kubilay Tanrikulu **3tl** SS/David Brimm **3br** SS/Edward A. Fink

Chapter 1: Newton, Motion, and Classical Mechanics
4 SS/Cody DeLong **5t** JI **5b** IS **6tl** SIL **6bl** IS/Lewis Wright **7tl** SS/Alex James Bramwell **7br** LOC **8tl** IS/David Elfstrom **8bl** SIL **9tr** PR/John Howard **10tl** IO/Vstock, LLC **10bl** AbleStock **12tl** SS/Dennis Sabo **12tr** JI **13bl** IS **14tl** SPL/NASA **14bl** JI **14tr** **15tl** **15r** SS/Jeff Thrower **16tl** IO/FogStock **16c** IO **17** SPL/TRL, Ltd. **18tl** **18r** NASA **18bl** SS/Tiburon Studios **19** JI **20tl** JI **20br** SS/Yuri Acurs **21tl** IS/Stan Rohrer **21tr** SS/ Suzanne Tucker **22tl** SPL/Andrew Lambert Photography **24tl** IS/HooRoo Graphics **24bl** SS/Andre Nantel **24tr** SS/Ulrike Hammerich **24tl** JI **24bl** JI **25tl** SS/VisualField **25tr** AP/Alberto Ramella **26tl** JI **26bl** SS/Ron Hilton **27** JI

Chapter 2: Physics and the Sky
28 JI **29t** SS/Risteski Goce **29b** IO/FogStock **30tl** SIL/ John Pendleton **30b** NGIC/P. Stattmayer **31tl** SPL/Mark Bond **31bl** NGIC/Kenneth Garrett **32tl** SS/Sebastian Laulitzki **32br** PR/Sheila Terry **33tl** PR/Mary Evans **33tc** SPL **33br** SPL/David A. Hardy **34tl** SIL **35** WI **36tl** SS/Marilyn Barbone **36b** SPL/David Hardy **37** SPL/ Mary Evans

Chapter 3: Waves
38 SS/Anette Linnea Rasmussen **39t** SS/Soundsnape **39b** JI **40tl** SPL/Andrew Lambert **41l** SPL/Andrew Lambert **41r** JI **42tl** SS/Eric Bechtold **42b** IS/Inozemtcev Konstantin **43** SPL **44tl** SS/Naomi Hasegawa **44b** JI **45bl** JI **45tr** SS/Eric Gevaert **46tl** JI **47** IS/Ted Denson **48tl** SS/Arlene Jean Gee **48cr** SPL/Edward Kinsman **48b** SPL/Andrew Lambert **49** SPL/Andrew Lambert

Chapter 4: Matter and Energy
50 SPL/Mike Agliolo **51t** IS/Andy Greene **51b** JI **52tl** SS/J. Helgason **52** SS/R **53tl** SS/Michael Thompson **53br** SPL/Sheila Terry **54tl** IS/Amanda Rohde **54bl** SS/Terrie L. Zeller **54tr** SIL/J. Caldwell **55** SIL/Cook **56tl** IS **56bl** SS/Popi Dimakou **56bc** SS/Milos Jokic **57** SS/Kaleb Timberlake **58tl** SS/SF Photography **58bl** JI **58bc** IO/Photolibrary.com **59** JI **60tl** JI **60bl** JI **61tr** PR/David R. Frazier **62tl** SPL/Andrew Lambert **62b** SPL **63cr** JI **63bl** SPL/Andrew Lambert **64tl** SS/Petur Asgeirsson **64l** SPL **65** SPL **66tl** IS **66b** SPL/Mark Burnett **67** SS/Peter Weber **68tl** JI **68bl** JI **68tr** SS/Eric Gustafson **69** JI

Chapter 5: Thermodynamics
70 NASA-MSFC **71t** JI **71b** SS/Dan Briski **72tl** SS/Merlin **72bl** SS/Roman Milert **73tl** SS/Natthawat Wongrat **73br** SS/J.T. Lewis **74tl** SS/Carsten Medom Madsen **76tl** SS/R **77tl** SS/Oystein Litleskare **77tr** PR/New York Public Library **78tl** SS/Robert Pernell **78b** LOC **79bl** LOC **79cr** SS/Kirsty Pargeter **80tl** SS/Suzanne Tucker **80b** SS/Mark Plumley **81tl** PR/SPL **81** PR/Science Source **82tl** SS/G. Tibbetts **82bl** SIL **82br** SIL **84tl** SS/Ron Hilton **84l** SIL **85** PR

Chapter 6: Light and Optics
86 SS/ David Brim **87t** SS/Michael Thompson **87b** JI **88tl** NASA **88tr** SIL **89t** SPL **90tl** SPL/Andrew Lambert **90br** SIL **92tl** SS/Peter Baxter **92bl** SS/Jostein Hauge **93t** SPL/Richard Menga **93br** SS/Elena Elisseeva **94tl** SS/Alan Heartfield **94bl** PR/Mary Evans **94tr** SS/Litwin Photography **95** PR/SPL **96tl** IS/Marcin Balcerzak **97t** SPL/David Parker **97bl** SPL/David Parker **98tl** SS/Marino **99** SS/Semen Lixodeev

Chapter 7: Electricity and Magnetism
100 JI **101t** SS/Nir Levy **101b** SS/Falk Kienas **102tl** IS/Kelly Borshiem **102bl** PR/David Taylor **103cl** PR/SPL **103br** IS/Arturo Limon **104tl** NASA-MSFC **105bl** WI **105tr** SPL/ Emilio Segre/ Visual Archives American Institute of Physics **106tl** SS/Arturo Limon **106tr** SS/Leonid Nishko **106br** SS/Dewayne Flowers **107** PR/Mark Burnett **108tl** SS **108bl** PR/ Sinclair Stammers **108br** SS/Thomas Mounsey **109** PR/Russ Lappa **110tl** SPL **110bl** LOC **110tr** IS **111** PR/NY Public Library **112tl** IS/Gregg Harris **112cl** PR/Sheila Terry **112br** PR/Andrew Lambert **114tl** SS/4uphoto.pt **114bl** SS/LaNae Chrsitenson **115bl** SS/Robert J. Beyers **116tl** SPL/Andrew Lambert **116bl** PR/SPL **117** IS/Roman Krochuk **118tl** SPL/Adam Hart-Davis **118cl** SIL-Dibner **119bl** SS/Alan Heartfield **119tr** SS/Chris Galbraith **120tl** JI **120br** SI/Laurie Minor-Penland **121cl** SPL **121tr** SPL **122tl** SS/Bill McKelvie **123bl** NASA KSC **123tr** SS/Cora Reed

Ready Reference
124l LOC **124tc** LOC **124bc** WI **124r** SS/Alfio Ferlito **125tl** LOC **125cl** SS/Kiyoshi Takahase Segundo **126** SXC **127** SXC **128cl** IO **128bl** SS/Amy Walters **128tr** IO/Photos.com Select **128br** SS/Manuel Fernandes **129t** JI **129c** SS/Falk Kienas **129b** IO/ImageDJ

Chapter 8: Radioactivity and Early Atomic Theory
130 SS/Anita **131t** IS/Brendon De Suza **131b** Matjaz Boncia **132tl** PR/Astrid & Hanns-Frieder Michler **132bl** PR/Library of Congress **132tr** PR **133** LOC **134tl** IS/Johanna Goodyear **134cl** SS/Robert Pernell **134bl** SS/Scott Rothstein **135** SPL **136tl** SPL/Frances Evelegh **136bl** LOC **137** SS/Ali Mazraie **138tl** NASA/Andrew Fruchter **139tr** SPL/Prof. Peter Fowler **139c** SPL/Frances Evelegh **140tl** PR/Jon Lomberg **141tr** SPL/James King-Holmes **141br** NASA

Chapter 9: Quantum Physics
142 SS/John Teate **143t** SS/Dario Diament **143b** SS/Johann Helgason **144l** SPL **146tl** SS/Jyothi Joshi **146bl** PR **147** SS/Handy Widiyanto **149tl** SPL **149br** SPL/Emilio Segre Visual Archives- AIP **150l** SPL **152tl** SPL/Andrew Lambert **152bl** SIL **153t** SPL/AIP **154tl** PR/AIP **154b** SPL/Volker Steger **155tl** SPL/Erik Heller **156tl** SS/Steve Simzer **157** SS/George Michael Warnock **158** SS/PhotosLB **159** SS/Scott Milless

Chapter 10: Einstein and Relativity
160 LOC **161t** JI **161b** SS/Alexis Puentes **162tl** LOC **162bl**LOC **162tr** SIL **163** PR/Sanford Roth **164tl** SIL **166tl** SS/Jenny Solomon **166b** NASA **167tr** SS/Graphyx **167br** PR/ Detlev van Ravenswaay **168** WI **169tr** NASA **169br** SS/Photomedia.com

Chapter 11: Fission, Fusion, and the Bomb
170 SS/Sebastian Kaulitzki **171t** IO/Photolibrary. com **171b** PR/AIP **172tl** PR/Michael Gilbert **173** SPL-AIP/University of Chicago **174tl** IO/AbleStock **176tl** PR/NASA

Chapter 12: The Standard Model
178 © 2006 Interactions.org **179t** Brookhaven National Laboratory Historic Image Library **179b** © 2006 Interactions.org **180tl** PR-AIP/Physics Today **181bl** PR/CERN **182tl** NASA **184tl** © 2006 Interactions.org **184b** © 2006 Interactions.org
185tl BNL

Chapter 13: Moguls of Modern Cosmology
186 NASA-JPL-Caltech/W. Reach **187t** NASA/JPL-Caltech **187b** NASA and The Hubble Heritage Team/C. Conselice University of Wisconsin STScl **188tl** NGIC/P. Stattmayer **188bl** AP **188tr** NASA-GSFC **189** NASA-GRIN **190tl** NASA **190bl** SPL-AIP/Emilio Segre Visual Archives **191**NASA-STScI **192tl** NASA-MSFC **192bl** NASA, ESA, H. Bond, STScI, & M. Barstow University of Leicester **193tl** LOC **193br** NASA, ESA, and H. Richer University of British Columbia **194tl** NASA **194br** PR-AIP/Emilio Serge Visual Archives **195tl** WI **195bl** NASA **196tl** NASA/JPL-Caltech **196bl** PR/Victor Habbick Visons **197** NASA-GSFC

Chapter 14: Open Questions and the Frontiers of Physics
198 NASA/JPL-Caltech **199t** NASA/JPL-Caltech/STScI/ Vassar **199b** Chandre-Harvard **200tl** NASA/L. Ferrarese of Johns Hopkins University **201** NASA, ESA, A. M. Koekemoer, STScI, M. Dickinson and The GOODS Team **202tl** Courtesy of SLAC and Nicolle Rager; © 2006 Interactions.org **202bl** CERN/interactions.org **203** NASA-GSFC **204tl** interactions.org/Sloan Digital Sky Survey **204tr** SPL/David Parker **204br** SLAC **205** interactions.org/Norman Graf **206tl** Courtesy of Fermilab Visual Media Services **206r** PR/Science Source **207tl** SPL/CERN 207br SPL/CERN **208tl** Courtesy of the Stanford Linear Accelerator Center **208bl** NASA/Les Bossinas **209t** Chandra-Harvard **209cr** interactions. com/Jean-Francois Colonna

Cover
Front SPL/Cordelia Molloy **Background** Jupiter Images/Steve Allen.